D0051918

THE COMPLETE

ENERGY-SAVING
HANDBOOK

FOR HOMEOWNERS

HARPER & ROW, PUBLISHERS

NEW YORK

Cambridge
Hagerstown
Philadelphia
San Francisco

London
Mexico City
São Paulo
Sydney

For Dwight Morrison

First Edition, 1979.

Portions of this book were published in government pamphlets entitled *In the Bank . . . Or Up the Chimney?*, *Solar Hot Water and Your Home*, *Project Retrotech*, *Home Heating in an Emergency*, *Save Energy, Save Money*, and *Design Ideas for the Energy Conscious Consumer*.

Library of Congress Catalog Card No.: 79-2808

ISBN: 0-06-465108-8

81 82 83 84 22 21 20 19 18 17 16 15 14 13

Printed in the United States of America

ACKNOWLEDGEMENTS

The facts and material in this book are based on the findings and methods published in these Federal Government reports: *Retrofitting Existing Housing for Energy Conservation: An Economic Analysis* and *Making the Most of Your Energy Dollars* by the Commerce Department's National Bureau of Standards, the Department of Energy's "Project Retrotech, Revised" the Department of Housing and Urban Development's *In the Bank . . . Or Up the Chimney?*, the Office of Economic Opportunity's *Save Energy: Save Money" Passive Design Ideas for the Energy Conscious Consumer* by the National Solar Heating and Cooling Information Center (U.S. Department of Housing and Urban Development in cooperation with U.S. Department of Energy), "Heating with Wood" from the Northeast Regional Agricultural Engineering Services of the Northeast Land Grant Universities and the U.S. Department of Agriculture, and "The Creosote Problem" from the Energy Saving Series (#3) of the New Hampshire Extension Service.

LEGAL NOTICE

CONTENTS

PREFACE

To American Homeowners:

Every homeowner seems to be worried about energy nowadays. There are some good reasons for this. The price of home heating fuels has doubled in the past two years and will go higher. For many people, this creates problems. Comfort and warmth are very important in the winter. Health depends on these—particularly the health of older people, small children, and those who are sick.

Higher heating costs have come at a time when other prices have gone up as well. Inflation has driven food prices up. Even the prices of heat-saving materials such as insulation, caulking, and weatherstripping are going up. Many people can benefit from weatherizing their homes, and the time to do it is NOW! Most materials are available at your local hardware store.

Energy-saving techniques applicable to those types were gathered and compared as to cost and potential fuel savings. The comparison was designed to emphasize safe and cost-saving energy conservation techniques that would return the greatest practical net savings to the homeowner over the life of the investment. This book will help you to grasp quickly the essential energy-saving techniques that your home may require and to choose wisely between do-it-yourself projects and contractor services.

This book is designed to take the confusion and guess-work out of home energy conservation practices. Using the home audit worksheets in the "Retrotech Job Book" you can determine the best combination of improvements for *your* house, climate, and fuel costs—improvements that will provide the largest long run savings in your home heating and cooling expenses.

Step-by-step installation instructions for each energy-saving method described are combined with illustrations that can take the mystery out of seemingly complex tasks. Detailed but easy forms enable you to compute the cost-savings of each method to help you save dollars and cents in home energy expenditures.

Homeowners, like other consumers, are often victims of a lack of information. It is sometimes difficult to determine how serious a problem a malfunctioning furnace, termites, or faulty plumbing might be—and consequently, how much a homeowner should invest to correct such problems.

In the same way, it is also difficult to determine how much insulation is cost-effective, how good an investment storm windows are, etc.

This Energy-Saving Handbook will help you determine what measures will make your home more energy efficient, and what you can expect to save in home heating and cooling costs by taking these measures.

Everyone's home is different. The simple procedures described in this workbook will allow you as much as possible to consider the unique features of your home. However, the savings you calculate may not be precise since many of the energy-saving measures are dependent upon how thermally efficient your home is already. Not all of these simple procedures take this into account. The calculations should, however, give you a good feel for the amount of money and energy you could save on your fuel bills.

Utility bills will continue to increase, but the rise in your personal bills will be less if you make judicious home improvements for energy conservation. You might think of such improvements as a hedge against inflation. The cost of making these improvements may well increase the value of your home. And, at current fuel prices, these improvements can pay for themselves many times over the life of the house.

This handbook is meant as a guide to show you what steps you can take to reduce your home energy costs, and how to carry out those steps that appear to be most cost-effective.

Household energy costs can be reduced with relatively simple home improvements. Installing or adding insulation, storm windows and doors, and applying weather stripping and caulking are important. But how much insulation is enough? When are storm windows and storm doors good investments?

Other sections of this handbook include: saving energy with "your heating, air conditioning, and water heating" systems, heating stoves and fireplaces, and more ways on how to save energy in the home. Special chapters focus on "passive solar energy and solar hot water systems."

You know that energy prices are still on the rise. That's why *now* is the time to invest in energy conserving measures that will help lower your costs in home heating and cooling operations. After all, saving energy means saving money. The final section of the book discusses how to obtain energy tax credits on next year's tax return.

If you are a homeowner who wants to save energy and money without sacrificing comfort, this book is for you. It is a "how-to" and a "how much" guide to energy conservation investments. Making the most of your energy dollars is the only way a money conscious homeowner can evaluate the costs and investments in order to offset rising energy prices. If you follow this guide you will be saving money and at the same time doing your part to help conserve our nation's precious energy supplies.

J.W.M.

SPECIAL TAXPAYER'S NOTE

Under the National Energy Act of 1978, taxpayers receive tax credits for installing energy-saving materials in their homes amounting to 15 percent of the first $2,000 spent on qualifying equipment, up to a maximum of $300. The credit can be subtracted from taxes due.

The credit could be applied to any equipment installed after April 20, 1977, the day President Carter announced his energy plan. The credit would be for existing dwellings only (in existence as of the effective date of the act). The credit would be effective through Dec. 31, 1985. The installation must be made in the taxpayer's principal residence. Eligible for the credit are owners, renters, and owners of cooperatives or condominiums.

A taxpayer who qualified for a tax credit in excess of the tax he owes could carry the credit forward on future tax returns through the taxable years ending Jan. 1, 1988. However, for expenditures made in 1977, a credit could only be claimed on the taxpayer's 1978 tax return. Be sure to check with the local office of the Internal Revenue Service before applying for the latest tax credit for energy saving improvements.

Section 1

HOME HEAT LOSS

A Quick Quiz

1. What is Your Thermostat Setting?   Score

If your thermostat is set at 68°F. or less during day-time in winter, score 6 points; 5 points for 69°; 4 points for 70°. If your thermostat is set above 70°, score 0.

_____

If you have whole-house air conditioning and you keep your temperature at 78°F. in the summer, score 5 points; 4 points for 77°; 3 points for 76°. If you have no air conditioning, score 7 points. If your thermostat is set below 76°, score 0.

_____

In winter, if you set your thermostat back to 60°F. or less at night, score 10 points; 9 points for 61°; 8 points for 62°; 7 points for 63°; 6 points for 64°; 5 points for 65°. If your thermostat is set above 65° at night, score 0.

_____

2. Is Your House Drafty?

To check for drafts, hold a flame (candle or match) about 1 inch from where windows and doors meet their frames.

If the flame doesn't move, there is no draft around your windows, and you score 10 points. If the flame moves, score 0.

_____

If there is no draft around your doors, add 5 points. If there is a draft, score 0.

_____

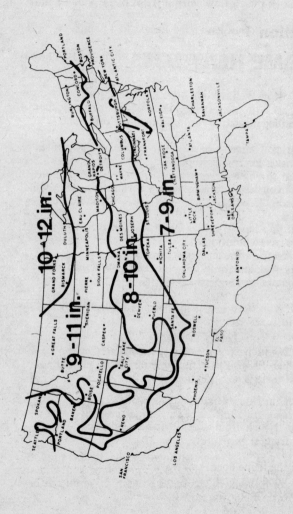

If you have a fireplace and keep the damper closed or block the air flow when it is not in use, add 4 points. _____

If you do not have a fireplace, add 4 points.   _____

If you leave the damper open when the fireplace is not being used, score 0.   _____

3. How Well is Your Attic Insulated?
Check the map (page 11) to determine the inches of ceiling insulation recommended for your zone.

If you already have the recommended thickness of insulation, score 30 points.   _____

If you have 2 inches less insulation than you should, score 25 points.   _____

If you have 4 inches less insulation than you should, score 15 points.   _____

If you have 6 inches less than you should, score 5 points.   _____

If you have less than 2 inches of insulation in your attic, score 0.   _____

4. Is Your Floor Insulated?
If you have unheated space under your house and there is insulation under your floor, add 10 points; if there is no insulation, score 0.   _____

If you have a heated or air condition basement or if there is no space under your house, score 10.   _____

5. Do You Have Storm Windows?
If you live in an area where the temperature frequently falls below 30°F. in winter and you use storm windows, score 20 points. If you do not have storm windows, score 0.   _____

Your Energy Quotient: Total _____
How Well Did You Do?

How Well Did You Do?

90 or above: Congratulations! You are already an energy saver. By keeping your home well-insulated and draft-free, you are using energy more efficiently than 80 percent of your neighbors, based on the national average.

Under 90: You're spending more money than you need to in order to keep your home comfortable. Check the quiz again to see where you lost the most points. That's where you can make the greatest savings in your annual fuel bill, while improving the comfort, appearance, and resale value of your home as well.

How Much Money Can You Save?

This workbook contains some simple calculations that will tell you approximately how much. These calculations are based on average housing, fuel, and climate conditions for your area. If your house is unusual in construction or location, you may want some additional advice before making your calculations. If so, please call your State Energy Office. People who are familiar with your area and conditions will be available to help you.

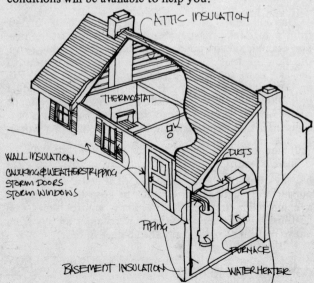

Heat Loss Calculation

Being comfortable in winter means keeping warm. This requires heat, which comes from increasingly expensive fuel. Most homes can use much less fuel without sacrificing comfort.

This manual explains the facts about winter comfort heating. It provides an easy method of approximating how much heat will be needed to keep any particular building warm and explains how to assess the benefits of improvements made to the building, such as adding storm windows, insulating exposed areas, excluding drafts.

Most such improvements cost money. Usually they save more than they cost. In the next few pages we shall see how to make heating improvements and how to figure the heat saved.

One important fact is that heat always tends to flow from a high-temperature area to a low-temperature area. For example, if you put a pan of cold water on a hot stove, the flow of heat from the stove through the bottom of the pan heats the water up to a higher temperature. Pan bottoms are, therefore, made of materials which conduct heat easily. To keep the pan from losing heat after it comes off the stove, you can stand it on an asbestos pad, a material that resists passage of heat or, in other words, provides insulation.

We put heat in a house to keep comfortable (and healthy) but the heat passes out of a house to the cold outside surroundings. If we want to keep the building at a comfortable temperature, we must control the actual heat-loss. Remember that the flow of air is always from hot to cold, and cold is really the absence of heat.

Begin your heat loss calculations with your ENERGY CONSUMPTION RECORD.

ENERGY CONSUMPTION RECORD

Year: _____

MONTH	ELECTRICITY		FUEL CHECK: GAS OIL COAL	
	KWH	COST	QUANTITY	COST
JANUARY				
FEBRUARY				
MARCH				
APRIL				
MAY				
JUNE				
JULY				
AUGUST				
SEPTEMBER				
OCTOBER				
NOVEMBER				
DECEMBER				
TOTAL PER YEAR				

ANNUAL ENERGY CONSUMPTION IN HEATING UNITS

QUANTITY				HEATING UNITS
1. _____ ELECTRICITY	KWH	÷	30	= _____
2. _____ NATURAL GAS	MCF*	÷	.12	= _____
3. _____ OTHER GAS**	GAL.	÷	1.3	= _____
4. _____ OIL	GAL.	÷	1	= _____
5. _____ COAL	LBS.	÷	15	= _____
6. TOTAL HEATING UNITS · · · · · · · · · · · ·				_____

*THOUSAND CUBIC FEET   **PROPANE, LPG OR BOTTLED GASS

USING THIS FORM PROVIDES THE NECESSARY INFORMATION FOR "YOUR HOME WEATHERIZATION," AND AS A BASIS FOR DEVELOPING CONTINUOUS RECORDS OF YOUR ENERGY USAGE!

Take the building below —

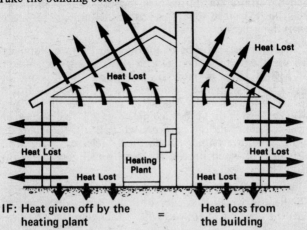

IF: Heat given off by the   =   Heat loss from
    heating plant   the building

THEN: TEMPERATURE INSIDE REMAINS CONSTANT

If a building has insulated walls, floors, and ceilings, double-glazed windows, and sealed cracks, then less heat will be needed to maintain comfortable temperatures inside the building. Less heat required means less fuel used, which means money saved.

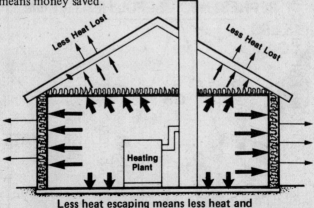

**Less heat escaping means less heat and
less fuel needed to stay comfortable inside**

Heat escapes from a building in two ways: by conduction

and by infiltration. In the next few pages, these processes will be explained—but first some definitions to make everything clear.

We must have a unit to measure heat losses. Normally we use the British Thermal Unit—Btu. This is the amount of heat it takes to raise the temperature of 1 pound of water by 1° Fahrenheit. Another way of "sizing" a Btu is to say it is about the amout of heat given off when a wooden match is burned completely. All fuel values or heat requirements can be expressed in Btu's: for example, using 1 kilowatt-hour of electricity releases 3,412 Btu; 1 pound of wood burned completely will give off about 8,000 Btu.

Working with Btu's means doing calculations with large numbers in which it is easy to make errors. This manual uses the concept of *Heating Units* to simplify the figuring—

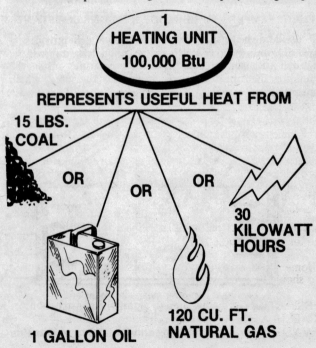

1 HEATING UNIT 100,000 Btu

REPRESENTS USEFUL HEAT FROM

15 LBS. COAL

OR   OR   OR

30 KILOWATT HOURS

1 GALLON OIL   120 CU. FT. NATURAL GAS

Obviously, the *heating unit* is an approximation because not all oil or gas heating plants operate at the same efficiency. However, the *heating unit* is a fairly accurate estimate of what a normal oil or gas furnace should get out of the quantities of those fuels illustrated above. The heating unit does, in fact, represent about 100,000 Btu.

In this manual, we calculate heating requirements on a seasonal basis. Therefore, if a building is calculated to require 1,200 *heating units,* that figure represents the approximate number of gallons of oil it should use per year, if it has an efficient furnace. If the actual fuel use is known and is very different from the calculated figure, a further check needs to be made to find the reason. It may be due to calculation errors, wrong measurements in the building, or a poorly functioning furnace. It may be that the building has only been partially heated previously, with much of the living space not used in the winter. Working with *heating units* in this way can enable us to spot errors or circumstances we might otherwise miss. Incidentally, if the furnace which provides heat for the building also provides domestic hot water, this will increase the fuel use approximately 20 percent.

To allow for climatic differences between areas, heating engineers use degree-day figures: 1 degree-day represents a 24-hour period in which the average outside temperature is 1°F below a base temperature of 65°F. Many northern areas will have over 7,000 degree-days in a heating season. This manual uses the *district heating factor* which, for an area having 4,000 heating degree-days, will be 1, for 6,000 degree-days 1.5 and so on. The DISTRICT HEATING FACTOR MAP shows *district heating factors* for various areas. Simply look up the approximate factor for your area, and use it in figuring all heat losses for any house in your district (see pages 28-29).

Home weatherization requires four steps to determine what should be done, where, and how:

1. INSPECTION of the building to determine construction.
2. CALCULATION of heat losses from the building.
3. EVALUATION of the building and heat losses to deter-

mine what weatherization measures should be made.
4. INSTALLATION of the weatherization materials.

A job book is used for recording the information on each building. It also shows the procedure for calculating and summarizing heat loss and serves as an order form for listing and procuring materials. (Pages 38-72).

Building Heat Loss by Conduction

Heat is lost from the home through the exterior surface of the building as heat flows by conduction through the building materials. The rate of heat loss from the warm side to the cold side through the exterior surface depends on the size of the surface, the length of time the heat flow occurs, the temperature difference between the two sides of the exposed area, and the construction of the section (the type of material used in the construction). All materials used in building construction reduce the flow of heat. Some materials are much better than others at reducing heat flow. The more effective materials are used as insulation. A well-insulated building will also stay cooler in the summer.

TYPES OF INSULATION

Three general types in insulation materials are commonly used in building construction. They are loose fill, blanket or batt, and rigid insulation.

Loose fill include such types of insulation as glass, rockwool, cellulose fibers, and wood fibers. Fill type insulating materials are best utilized on horizontal surfaces, such as ceiling areas. This type of insulation used in vertical areas tends to settle, and unless provision is made to refill the space, cold spots can occur.

Blanket, or **batt insulation,** is commonly made of glass, rockwool, or wood fiber. They are usually enclosed in a paper envelope or fastened to a backing of kraft paper or aluminum foil. Some blanket types of insulation have no backing and are intended to be used when no vapor barrier is required. Blanket insulation comes in rolls of various lengths and thicknesses. Batt insulation is usually thicker and comes in shorter lengths. Both blanket and batt insulation are available for framing spacing of 16 and 24 inches. Other widths are available on special order.

Rigid insulation, in addition to providing insulating value, also provides structural strength. Rigid insulation is available in board form, such as various fiberboard materials and foamed plastics. Rigid insulation is used quite extensively by contractors and not individual homeowners. In some instances, this type insulation is less expensive.

Table 1 (page 32) lists the insulating value of most of the material found in house construction. The R value shown in the right-hand column indicates the effectiveness, or resistance value of the material. *The higher the resistance value, the better the insulating quality.* When building sections are made of several materials, the resistance value of each of the individual materials can be added together to obtain the overall total resistance value. Once you know the overall R value you can use it in the calculation outlined in the job book to determine heat loss. Thicknesses of 3½ inches (R-11) and 6 inches (R-19) are most common.

Vapor Barriers

In the winter, moisture moves from the inside of the home to the outside through the exterior surfaces. Vapor barriers are installed to reduce the flow of moisture through

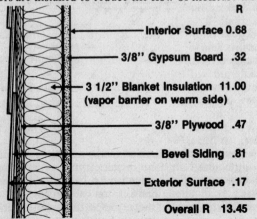

	R
Interior Surface	0.68
3/8" Gypsum Board	.32
3 1/2" Blanket Insulation (vapor barrier on warm side)	11.00
3/8" Plywood	.47
Bevel Siding	.81
Exterior Surface	.17
Overall R	**13.45**

the insulation so that condensation will not occur. Blanket or batt insulation usually has vapor barriers attached. Polyethylene film (4 mils thick) can be used as a separate vapor barrier if needed. Vapor barriers should always be installed on the warm side (inside) to stop the moisture before it reaches the insulation. If possible, vent the cold side of the insulation to the outside to remove moisture which escapes through the insulation. When a blanket or batt insulation having an attached vapor barrier is used, kraft paper backing is usually cheaper than foil backing. If foil backing is used, a ¾" to 3" clearance is necessary in order to maintain effectiveness.

Exterior Walls

To determine the insulating value of an exterior wall section, it is necessary to know the construction of the wall. Using Table 1, determine the R value for each material making up the wall. Add together these values to obtain the overall R value of the wall.

Ceilings and Roofs

The insulating value of roof and ceiling sections can be determined by adding the R value of each of the materials making up the section. It is necessary to know the construction of the ceiling or roof section. Add together the R values of the materials making up the section from the values given in Table

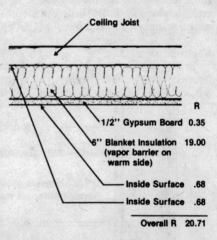

	R
1/2" Gypsum Board	0.35
6" Blanket Insulation (vapor barrier on warm side)	19.00
Inside Surface	.68
Inside Surface	.68
Overall R	**20.71**

1 to obtain the overall R value of the section. For ceilings having attic space over the insulation, use an interior surface resistance for the surface next to the attic due to the fact that still air conditions exist on the outside of the insulation. The illustration shows the procedure for determin-

ing the overall resistance value for ceiling sections.

For roof sections, the procedure to determine overall resistance value is similar to that for the wall section. First, determine the construction of the roof section, and then add the resistance of the individual materials making up the roof section to obtain overall R value for the roof. The diagram at the right shows the procedure for determining the overall R value for roof areas.

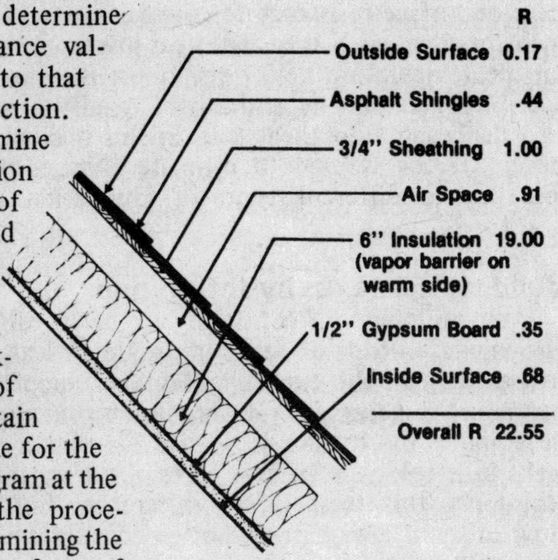

	R
Outside Surface	0.17
Asphalt Shingles	.44
3/4" Sheathing	1.00
Air Space	.91
6" Insulation (vapor barrier on warm side)	19.00
1/2" Gypsum Board	.35
Inside Surface	.68
Overall R	**22.55**

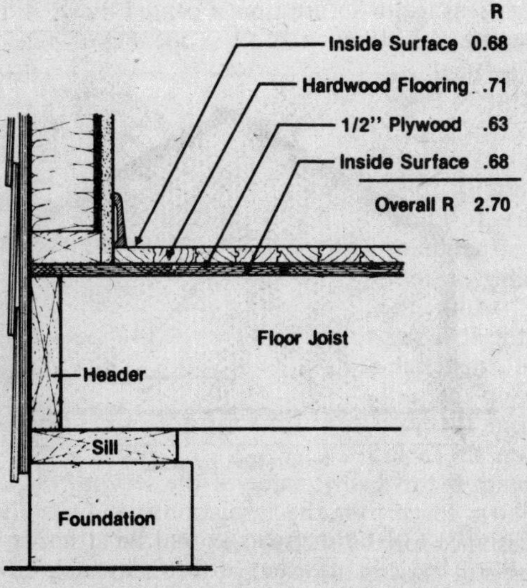

	R
Inside Surface	0.68
Hardwood Flooring	.71
1/2" Plywood	.63
Inside Surface	.68
Overall R	**2.70**

Floor Joist

Header

Sill

Foundation

Floors

To determine the insulating value of floors, add the R

value of the individual materials making up the floor section together to determine the overall R value. Use the interior surface resistance for the surface next to the basement or crawl space area. The heat loss from floors depends on the temperature below the floor. Basement and crawl space temperatures depend on the quality of construction. In calculating floor heat loss in this manual, a floor exposure factor is used to estimate changes in floor heat loss due to different types of foundation construction.

Building Heat Loss by Infiltration

Any building will constantly exchange air with its environment: outside air leaks in, inside air leaks out. A certain amount of this exchange (say, one complete air change per hour) is necessary for ventilation, but most buildings have much more than is needed. In winter, the air that leaks in is cold; the air that leaks out is warm; fuel is used to supply this temperature difference. Exfiltration (the flow of air is always from hot to cold) is the main reason for a cold house.

This leakage or infiltration is caused by wind, the building acting as a chimney and the opening of outside doors.

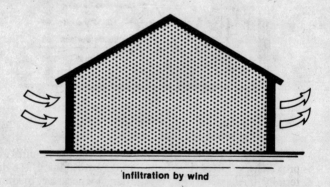

infiltration by wind

The effect of door openings and wind needs little explanation, but the chimney effect may not be obvious. When air in a building is warmer than the outside air, the entire building acts like a chimney—hot air tends to rise and leak out of cracks at the upper levels and sucks cold air in through cracks at the lower levels. Both the temperature difference and building height contribute to this effect. A two-story house having a 68°F inside temperature and a 30°F outside temperature will produce a "chimney" leakage equivalent to a 10 mile per hour wind blowing against the building.

EXFILTRATION

HOT AIR OUT

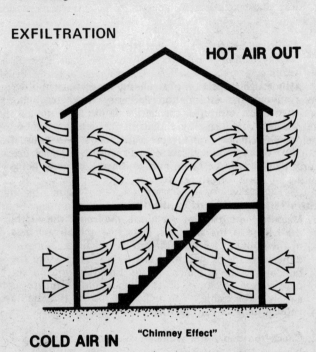

"Chimney Effect"

COLD AIR IN

Each cubic foot of air that enters the house requires approximately 0.02 Btu to raise the temperature 1°F. To determine the heat loss from infiltration, it is necessary to know the rate of air movement through the home. Most houses undergo from one to three air changes per hour, depending on construction. An infiltration checklist (page 31) is provided to determine the approximate infiltration rate.

Section 2

HOME ENERGY AUDIT

A thorough inspection of the house is necessary in order to find out the information necessary for determining weatherization measures. Check the house as thoroughly as possible, and make sure that all the details described below are collected and recorded. If you find unusual building construction, location, arrangement, or other features that need to be taken into consideration, make notes in the job book for future reference.

Step 1: Inspection of Building

1. **Measure the outside of the house, the length, the width, the height of the sidewalls.** Draw a sketch of each side of the house and a floor plan in the job book. Note the dimensions. It is not necessary that you measure the building to the last inch. If you find that you do not have access to some of the dimensions, estimate as closely as possible. You will find that it is accurate enough for your calculations.

2. **Check the doors and windows.** Sketch them in the job book in their proper location; note their construction, using an **S** for a single window or door and **D** for a double window or door. If you do not have access to a window, estimate its dimensions. Otherwise, measure and note the sizes in the job book.

3. **Check the construction of the exterior of the home.** Check the construction of the walls, the exterior ceiling, and roof and floor surfaces. It is necessary to know what

kind of materials make up these areas in order to determine how well they are insulated. If you cannot readily determine this, you may be able to find out by taking off a switch plate or a plate over an outlet box on the exterior wall to examine the inside of the wall. However, if you have difficulty finding what is in the wall or ceiling, chances are that you cannot do anything about it. Once you have found the construction materials for the exterior surface, list the construction materials in the job book under "Walls, ceiling and roof, or floors." Note the material and thickness of the material so that you can later determine its insulating value. Note also if you can get access to the area where the insulation will have to be installed. If, for instance, a wall is enclosed with sheathing material, it is difficult to add insulation. However, if you have an attic space over an assessible ceiling, you can insulate the ceiling.

4. **Check the building foundation to determine the floor exposure factor.** Put a checkmark in the box opposite the foundation description that best describes the actual building foundation.

5. **Check the condition of the home to determine the approximate infiltration rate which is part of the "General Waste of Heat" category. See separate Job Order Sheet.** Put a checkmark in the box under category 1, 2, or 3 that best describes the condition of each building component in the table on page 62 of the job book. In order to select the appropriate category, check for the following:

Building Foundation

To determine leakage around sills and cellar windows, examine the structure from the inside. Look for daylight between the sill and the foundation. Feel for drafts at cellar windows. Push on windows to see if they are loose or rattle. Check for missing putty on sash. This detailed analysis will help decide if the cellar is in category 1, 2, or 3.

If the building is on posts, infiltration must be eval-

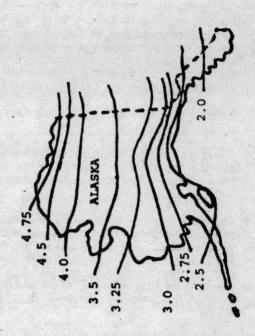

DISTRICT HEATING FACTOR
FOR THE
UNITED STATES

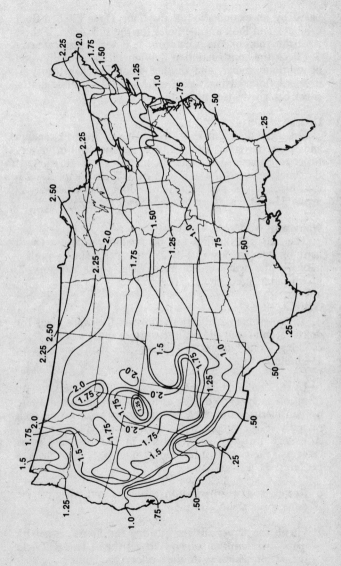

uated by an examination of the floor. Open kitchen floor cabinets, and look at pipe holes for the sink drain. If these are tight, the building is category 1; if very open, category 3. Check the construction of the floor: if made of plywood or subfloor, paper, and finished floor, check category 1; if board floor having visible cracks and discernible drafts, check category 3 (see page 31).

Doors

Open the door quickly—a good fit will create a vacuum and resist the effort to open. A loose door will offer very little resistance. An on-hands-and-knees examination of the crack between the door and sill will also help. If a 25-cent piece can be pushed under the door, check category 2; if two 25-cent pieces can be used, check category 3.

Windows

The same evaluation can be used on windows (push hard on the window): if a 25-cent piece can be pushed between the window and the casing, check category 2; if two can be used, check category 3.

Walls

To determine infiltration through the walls, feel for drafts around outside wall electrical outlets. Check for caulking around doors and windows, condition of paint, and building paper.

Usually building components are not all in the same infiltration category. You can estimate the approximate rate by considering how many of the components are in each category. For example, if two components are in the three air-change category and two are in the two air-change category, the overall infiltration would be 2½ air changes per hour.

6. Record the weatherization problems and what can be done about them.

7. Fill in the directions for locating the house, either by map or description of how to get there. This will help the person who is to do the work on the house.

The following are typical features of buildings having infiltration rates of approximately 1, 2, or 3 air changes per hour:

Building Component	One Air Change Per Hour	Two Air Changes Per Hour	Three Air Changes Per Hour
Building with cellar OR	Tight, no cracks, caulked sills, sealed cellar windows, no grade entrance leaks	Some foundation cracks, no weatherstripping on cellar windows, grade entrance not tight	Stone foundation, considerable leakage area, poor seal around grade entrance
Building with crawl space or on posts	Plywood floor, no trap door leaks, no leaks around water, sewer and electrical openings	Tongue and groove board floor, reasonable fit on trap doors, around pipes, etc.	Board floor, loose fit around pipes, etc.
Windows	Storm windows with good fit	No storm windows, good fit on regular windows	No storm windows, loose fit on regular windows
Doors	Good fit on storm doors	Loose storm doors, poor fit on side door	No storm doors, loose fit on inside door
Walls	Caulked windows and doors, building paper used under siding	Caulking in poor repair	No indication of building paper, evident cracks around door and window frame

TABLE 1: INSULATION VALUE OF COMMON MATERIALS

MATERIAL	THICKNESS IN INCHES	R VALUE
Air Film and Spaces:		
Air space, bounded by ordinary materials	¾ or more	.91
Air space, bounded by aluminum foil	¾ or more	2.17
Exterior surface resistance	—	.17
Interior surface resistance	—	.68
Masonry:		
Sand and gravel concrete block	8	1.11
	12	1.28
Lightweight concrete block	8	2.00
	12	2.13
Face brick	4	.44
Concrete cast in place	8	.64
Building Materials — General:		
Wood sheathing or subfloor	3/4	1.00
Fiber board insulating sheathing	3/4	2.10
Plywood	5/8	.79
	1/2	.63
	3/8	.47
Bevel-lapped siding	1/2 x 8	.81

Vertical tongue and groove board	3/4	1.00
Drop siding	3/4	.94
Asbestos board	1/4	.13
3/8" gypsum lath and 3/8" plaster	3/4	.42
Gypsum board (sheet rock)	3/8	.32
Interior plywood panel	1/4	.31
Building paper	—	.06
Vapor barrier	—	.00
Wood shingles	—	.87
Asphalt shingles	—	.44
Linoleum	—	.08
Carpet with fiber pad	—	2.08
Hardwood floor	—	.71

Insulation Materials (mineral wool, glass wool, wood wool):

Blanket or batts	1	3.70
	3 1/2	11.00
	6	19.00
Loose fill	1	3.33
Rigid insulation board (sheathing)	3/4	2.10

Windows and Doors:

Single window	—	approx. 1.00
Double window	—	approx. 2.00
Exterior door	—	approx. 2.00

Step 2: Calculation of Heat Loss

The next step in the home weatherization procedure is to calculate the heat losses expected from the house. Only simple calculations are necessary, and you can do it away from the job site. By following the job book step by step, you can calculate the amount of heat lost in a given season from the building.

Step 3: Evaluation of Data

Determining Weatherization Measures

In order to determine which weatherization measures should be undertaken on the home, we need to consider several factors. By following the step-by-step procedure listed below, you should be able to determine the logical areas where weatherization should be undertaken.

1. **Check the heat units** now required on column 5 of the summary table on page 65. Look for areas that have the highest heat requirement as logical areas for weatherization.

2. **Evaluate the comments** on page 67. The following table on page 36 may be helpful in determining what weatherization measures are indicated from these comments.

3. **All windows should be fitted with storm windows.** Since this measure increases the insulation value of the window, as well as reduces the infiltration through windows, it usually results in the greatest benefit. Storm doors do not give as much benefit as storm windows. Weatherstripping exterior doors is nearly as effective as storm doors and not as expensive.

4. **Check to see if insulation can be applied.** If there is not at least 6 inches of insulation in the ceiling, 3½ inches in the sidewall, and 3½ inches under the floor, and if the basement or crawl space is unheated, consider adding insulation. It usually will result in a substantial heat saving. Check the job book to determine if insulation can be added to the areas where it is needed. It is usually easy to add insulation to an attic space and difficult to add insulation to a sheathed wall or below the floor.

5. **Check to see if the floor exposure factor and infiltration from the foundation can be improved by adding banking materials.** If the floor is over an unheated area, consider insulating the floor; if over a crawl space, consider insulating the exposed foundation wall.

6. **Check to see if weatherstripping can be installed around single doors and windows or storm doors to reduce the infiltration rate.** Usually weatherstripping is not necessary if storm windows have been applied. This measure usually is not as effective in reducing heat loss as those listed above, but if for any reason you cannot install storm doors and windows, make sure the weatherstripping is adequate.

7. **Check to see if the infiltration through the wall can be reduced by caulking around doors and window frames.** This is one of the less effective measures of reducing heat loss; however, it may help in a loosely constructed building.

8. **You should now be able to determine where the most logical areas for weatherization should occur.** Fill in the proposed changes beginning on page 64 of the summary table in the job book, and calculate the heat saving from the proposed changes; use the procedure that was originally used in determining the heat requirements.

9. **Check the availability and price of weatherization materials in your area.** Visit several building supply dealers, and determine what kind of weatherization materials are available and the price. It is a good idea to make a list of these as you visit the dealer, and from these you will be able to select which dealer has the proper materials available at the most attractive price.

10. **Calculate the quantities of materials needed to complete the home weatherization in the job book.**

11. **Complete the job book;** list the materials required and their cost along with the instructions for applying the

Problem	Probable Cause	Remedy
Low house temperature High fuel use	High heat loss	Add insulation Add storm doors and windows Caulk and weatherstrip doors and windows
Cold Floors	Cold crawl space or basement	Add banking to increase crawl space or basement temperature
Drafty house. Results from convection currents	Loose doors and windows	Add storm doors & windows and caulk around windows and doors
Wet windows	Cold window surface or high humidity	Add storm windows or ventilate to reduce humidity
Wet walls or ceiling	Cold inside surface or high humidity	Ventilate to reduce humidity

materials. This sheet is to go to the job site when the
work is to be done, and any instructions required for
the installation of the materials should be made on this
sheet.

12. **Calculate the total cost of the weatherization materials**
on pages 64-72 of the job book, and calculate the payoff
time for weatherization materials.

☐	**X** ☐	**=** ☐

Fuel Factor
Fuel Oil = 1
Electricity = 30
Natural gas = 120, Wood = .01

Price of Fuel,
per gal.,
cu. ft., KWH, cord

**COST OF ONE
HEATING UNIT:**

PAY-OFF TIME OF ONE WEATHERIZATION ACTIVITY:
This is the number of seasons for fuel savings
to pay off the cost of this activity.

ACTIVITY: _____

$$\frac{\boxed{}}{\boxed{} \quad \text{X} \quad \boxed{}} = \boxed{}$$

Cost of activity
(from job sheet above)

Heating units
saved

Cost of one
heating unit
(from job sheet
above)

"Pay-off" time
(heating seasons)

Step 4: Completion of Home Audit Job Book

Before proceeding to Step 4: Special instructions for completing the job book are given below and on each page of the job book.

YOUR
HOME WEATHERIZATION
AUDIT

This revised Job Book for Home Weatherization Audit replaces the original Retrotech Job Book and is used in much the same way. Additional information is developed which allows priority of various weatherization measures for a building to be determined as required by the National Energy Conservation Policy Act.

This section is also intended to be used for determining costs of weatherization measures and consequent fuel savings for a particular building in an area, e.g. ranch houses, colonials, row houses, etc. to establish a priority listing which will hold for all similarly constructed buildings in the area.

This revised "Retrofit" Job Book (PART I) specifies the steps that must be followed for a complete home weatherization audit.

By using this job book it is possible to fill out the Building Check and Job Order Sheet (PART II) in performing an on-site inspection to determine which measures to use to weatherize a dwelling unit and to identify the weatherization materials and work to be undertaken. If the design of a dwelling unit makes the use of the list inappropriate, it is possible to complete the calculations in the Home Audit—Job Book to determine the most cost-effective set of measures to be used to weatherize the dwelling unit.

PART I
CALCULATION PROCEDURES FOR
WEATHERIZATION MATERIALS

There are six calculation procedures to be followed:
a. Description of Building
b. Heat Losses by Conduction through Uninsulated Ceilings
c. Heat Losses by Conduction through Partially Insulated Ceilings
d. Heat Losses by Conduction through Floors
e. Heat Losses by Conduction through Uninsulated Walls
f. Heat Losses by Conduction and Infiltration through Single-Glass Windows
g. Optional — heat loss by Conduction and Infiltration through Double-Glass or Plastic-Covered Windows and through Doors

(1) *Description of the Building* (pages 46-49)
  a. Enter into the form sketches of the various views of the building.
  b. Enter in the appropriate District Heating Factor from the map. If the type of the building is to be applied to an area containing several District Heating Factors, use the average of the factors. If this variation in heating factors would result in different rankings, division of the building type into two or more categories covering different areas would be appropriate.
  c. Enter in fuel prices per unit of fuel. The price should be a reasonable approximation of the price presently charged to occupants of dwellings in the category.

(2) *Heat Losses by Conduction through Uninsulated Ceilings.* (pages 50-51).

  a. From Table 1 of the Job Book, obtain the R value for each material in the ceiling (or roof, where appropriate). Add these R values to obtain a composite R value for ceiling (or roof). Round off the value to the nearest tenth of unit.
  b. Enter ceiling area and District Heating Factor and determine the heating units now required due to losses through the ceiling.

c. In a similar manner, determine the heating units which would be required if the ceiling were insulated to have an R value appropriate to the area.

The appropriate amount of insulation for a ceiling depends on climate, fuel prices and insulation costs. Check the current DOE recommendations on maximum R value for the area. These R values will be reviewed each year to keep the recommendations up to date.

d. Subtract to determine the potential heating savings and enter at the bottom of the page.

(3) *Heat Losses by Conduction through Partially Insulated Ceilings* (pages 52-53).

Some ceilings, particularly in Northern Tier areas, may have had some insulation installed when the house was built or some may have been added since. Though the typical building which is evaluated may not have any ceiling insulation, some houses of the same type may. In cases like this, although the primary savings have been made, it may be worthwhile adding more insulation. This page is used to determine if this is so.

a. Determine the heating units required by a ceiling like the one in the typical house if it were insulated to R-10 (that is if it had approximately 2-3 inches of insulation).

b. In a similar manner, determine the heating units which would be required if the ceiling were insulated to have an R value appropriate to the area.

c. Subtract to determine the potential heating savings and enter at the bottom of the page.

(4) *Heat Losses by Conduction through Floors* (pgs. 54-55).

a. Enter the floor area, appropriate exposure factor, District Heating Factor and R value and determine the heating units now required due to losses through the floor.

b. In a similar manner check the potential heating units based on a floor exposure factor of 0.5, provided draft proofing below the floor is practical.

c. If insulating the perimeter of the crawl space or basement is practical, determine the potential heating units if the floor exposure factor were reduced to

0.3 in this way.

d. If insulating the perimeter is not practical due to probable wetting of the insulation or need for under-floor ventilation, insulating the whole floor may be the only possible treatment. Determine the potential heating units with a minimum floor R-15.

e. Subtract the three different potential heating unit figures from the heating units now required and enter the figures in the boxes at the bottom of the page.

(5) *Heat Losses by Conduction through Uninsulated Walls* (pages 56-57)

If walls are already partially insulated, e.g. with a 2-inch blanket of insulation in a 4-inch cavity, it is usually impossible to add insulation to the wall. Even if the typical house has partial insulation this page might be completed in case some examples of the type are completely uninsulated.

At the present time, only frame walls are normally being insulated in the weatherization program, as a suitable treatment for masonry and log walls is not generally available. However if a treatment for such walls becomes available in the future, it can be evaluated by the procedure on this page. If the R value attained is other than 15, that value should be used in the potential savings calculation.

a. For uninsulated frame walls, a total R value around 3.0 can be used, no matter what the type of surface treatment. However if the walls are of different construction, it may be necessary to determine the R value. From Table 1 in the Job Book, obtain the R value for each material in the walls. Add these R values to obtain a composite value for the wall. Round off to nearest tenth of unit.

b. Enter perimeter and height of outside walls. Multiply to obtain gross wall area. Subtract the sum of areas of windows and doors, to determine net wall area.

c. Determine the heating units now required due to losses through the walls.

d. In a similar manner determine the heating units which would be required if the walls were insulated to R-15.

e. Subtract to determine the potential heating savings and enter at the bottom of the page.

(6) *Heat Losses by Conduction and Infiltration through Single Glass Windows* (pages 58-59)

Very often a significant infiltration loss occurs *around* windows in addition to the losses by conduction *through* the glass. Adding storm windows can significantly reduce infiltration losses as well as conduction losses. It is important to allow for this when the benefits of storm windows are being considered relative to other possible measures.

    a. Use the building sketches to fill in the table to determine the total area of single glass windows.

    b. Determine the heating units required for conduction losses through the windows (based on R-1 for the windows).

    c. Determine the potential saving by adding storm windows. The calculation allows for the storm windows saving half the conduction loss plus a saving of infiltration heat loss equivalent to one quarter of the loss through single glass.

    d. Subtract to determine the potential heating savings and enter at the bottom of the page.

(7) *Heat Losses by Conduction and Infiltration through Double-Glass or Plastic-Covered Windows and Through Doors (Other-optional)* (pages 60-61)

Some additional heat loss occurs around glass doors and/or plastic-covered windows. Triple glazing of windows can be done but is not usually practical. If no change were made in the windows, the potential saving would be 0 heating units. If windows were triple glazed, the R value would be approximately 3, and the potential savings would be one-third of the required heating units. The potential heating savings, if any, on the Summary Table for Priority Rating of Weatherization Measures under "other."

(8) *Summary Table for Priority Rating of Weatherization Measures* (pages 64-65)

This table allows the relative costs and savings for various possible weatherization measures to be compared. The most cost-effective measures can then be determined:

    a. Enter the potential heating savings from the boxes at the bottom of the pages on the appropriate line in column 5.

b. Estimate the cost of materials for each possible weatherization measure in Column 1, entering the unit cost ($cost/ft^2$ or other unit) of the materials on which the estimate is based. The unit cost should be an approximation of the average cost across the geographical area covered by the building type.

c. Enter the installation factor for each measure in Column 2. The installation factor is equal to the total cost (installation and material) of the measure installed divided by the material cost. Installation cost includes on-site labor and equipment rental. Standard reference guides which list installation costs can be used. The installation factor allows for the additional cost of installing the materials so a true cost of the weatherization measure can be obtained.

d. Multiply the total material cost by the installation factor for each weatherization measure to find Total Cost and enter in Column 3.

e. Enter the expected life of each proposed weatherization measure in Column 4 and enter the cost per heating unit for the fuel actually being used in Column 6. Then determine the lifetime savings due to each proposed weatherization measure by multiplying the figure in Column 4 by the figure in Column 5 and the product by the figure in Column 6. Enter the answers in Column 7.

f. The Cost:Benefit ratio for each weatherization measure is then calculated by dividing the figure in Column 3 (total cost of measure) by the figure in Column 7 (savings due to the measure) and is entered in Column 8.

g. Rank the cost-effectiveness of all the weatherization measures by comparing the figures for Cost:Benefit Ratio saved. Reducing general heat waste is always regarded as cost effective and is already entered as No. 1 in Column 9. The weatherization measure calculated in column 8 to have the lowest Cost:Benefit ratio (i.e. to cost the *least* per heating unit saved) is then entered as No. 2 in column 9. The next least costly measure is No. 3, and so on. Any measure which has a Cost:Benefit ratio exceeding 1.0 will probably cost more than it saves over its lifetime and should be evaluated very carefully at the local level before being included on the Recommended List.

(9) *Recommended Weatherization Priority List* (page 66)

Transfer the information on source of heat loss, appropriate weatherization measures and unit material costs for cost-effective measures *in priority order* from the Summary Table onto the Recommended Weatherization Priority List on the lower half of the page. Enter the building type and area covered (e.g. statewide or subarea by name) in the appropriate spaces.

(10) *Calculation of priority ranking by category*

a. Each State will use the Recommended Weatherization Priority List for inclusion in the State Plan. The State is to combine the ranking from the sample buildings in each category and enter the final ranking into a "Recommended Weatherization Priority List."

b. Enter the type of building and geographical area covered by the type in the appropriate spaces.

c. Some measures may be impossible to implement because of local conditions. For example, carrying out certain measures may violate State law. Any of the considered measures which are suitable for the local situation should be checked in the last column of the Recommended Weatherization Priority List. A notation of the reason for exclusion, e.g. "not cost-effective" or "unsuitable" should be made for any measure which is to be omitted from the Retrotech Building Check and Job Order Sheet.

PART II
BUILDING CHECK AND JOB ORDER SHEET
(see pages 67-73)

(1) Enter the estimated quantity of materials and cost in the spaces provided.

(2) Determine which measures can be installed within the cost guidelines of the program as provided by the State. Priority will be given to measures which reduce general heat waste.

(3) The Sheet can be used as an ordering form.

(4) Enter the actual expenditures for measures installed.

(5) Complete Job Order Sheet regarding General Waste of Heat, including Heat Loss by Infiltration.

Alternative procedure

You can elect to evaluate the dwelling unit (without using the State priority ranking of weatherization measures). This will typically occur where the dwelling unit does not fit any of the State building types or where the use of a type's ranking would result in an inappropriate set of measures. Using this alternative procedure, the following steps will be used:

(1) Conduct an evaluation of the dwelling unit using the Job Book calculation procedure and instructions found in Part I of the Job Book.

(2) Since Part I instructions were written for State use for calculating ranking by types, the program operator should use the following special instructions where appropriate.

   a. Fill out the form only for the dwelling unit to be weatherized.

   b. In the Description of Building Table use current fuel prices for the specific dwelling unit or geographic area.

   c. In the Description of Building Table complete the dwelling unit information consisting of name and address of owner, type of fuel used, etc.

   d. In the Summary Table use unit cost of material based upon local prices of such materials.

   e. In the Summary Table use either the State installation factors from the type of buildings most similar to the dwelling unit or estimate the factors using local data on on-site labor and rental equipment costs.

   f. Calculate either "Heat Losses by Conduction through Uninsulated Ceilings" or "Heat Losses by Conduction through Partially Insulated Ceilings" whichever is appropriate to the dwelling unit. If using the partially uninsulated table, calculate the actual Total R value instead of using the value "10."

(3) Place the ranked measure on the first page of the Building Check and Job Order Sheet.

(4) Complete the Building Check and Job Order Sheet in accordance with above instructions of this document.

Home
Weatherization
Job Book for:

Sketch all views and put dimensions on each part shown, for example, length of walls, width and length of windows, etc. Label all single glass windows S and double glass and doors D . Complete all items in the job book labeled "Fill in at Job Site."

Description of Building

Front View	Right Side View

District Heating
Factor

Rear View

Left Side View

Plan View

Calculation of Floor Area

If the building is not rectangular or square add the areas
of the various parts together to find the floor area

Building
Length
ft.

Building
width
ft.

Floor Area
sq. ft.

Description of Building

Dwelling Unit Information

Name of Head of Household _____

Style of Structure:

☐ One-story

☐ Two-story

☐ 2½ story

☐ Split-level

☐ Other _____
(specify)

Occupants of Structure:

☐ Total number _____

Heating System Information

Type of Fuel: (P = Primary)
(S = Secondary)

			COST PER HEATING UNIT ($)
☐ OIL	($ per gallon)	× 1 =	
☐ NATURAL GAS	($ per therm)	× 1.2 =	
		or	
☐ PROPANE	($ per cu ft)	× 120 =	
☐ ELECTRICITY	($ per gal)*	× 1.4 =	
☐ COAL	($ per kwh)	× 30 =	
☐ WOOD	($ per ton)	× .005 =	
	($ per cord)	× .007 =	

*($ per lb) × 6

Type of Heating System: (P = Primary)
(S = Secondary)

Age of Structure (approx.): _____ years

☐ Steam/hot water/hot air

☐ Fireplace/stove/portable heater

☐ Electrical baseboard

☐ Other (specify) _____

Rooms in Living Space:

Domestic Hot Water:

Does central heating system provide heat for domestic hot water?

☐ Total number of rooms

☐ Number used in winter

Thermostat Setting in Winter (average):

_____ Day _____ Night _____ None

Amount of Fuel Used Last Heating Season:

	Primary	Secondary
Type	_____	_____
Quantity	_____	_____
Total Cost	$ _____	$ _____

Occupants Comments (drafty, cold floors, too expensive to heat):

_____
_____
_____
_____
_____
_____

FILL IN
AT
JOB SITE

HEAT LOSSES BY CONDUCTION THROUGH UNINSULATED CEILINGS

Material	Thickness (inches)	R Value
Inside surface	–	0.68
Inside surface (0.68) — OR —	–	
Outside surface (0.17)	–	
	Total R value	

FILL IN AT JOB SITE

Area of Ceiling
(Take area of upstairs ceiling in a two-story house)

Ceiling area will normally be the same as floor area (from building description sheet)

[ ]

Distance between joists/rafters:

Always include the materials in the roof deck in addition to the materials in the actual ceiling.

[ ] Ceiling area sq.ft. **×** [ ] District heating factor **÷** [ ] Total R value **=** [ ] Heating units required

Potential Savings by Insulation of Ceilings

The appropriate amount of insulation for a ceiling depends on climate, fuel prices and insulation costs. Check the current DOE recommendations on maximum appropriate R value for your area. This indicates the maximum amount of insulation which is cost effective for your climate at current prices. Any less amount of insulation than this will save less total fuel but will save more per dollar spent.

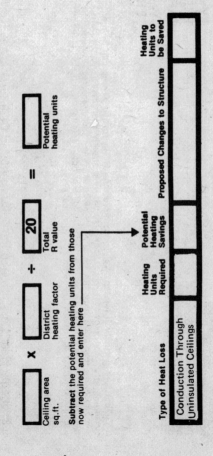

Subtract the potential heating units from those now required and enter here

Type of Heat Loss	Heating Units Required	Potential Heating Savings	Proposed Changes to Structure	Heating Units to be Saved
Conduction Through Uninsulated Ceilings				

HEAT LOSSES BY CONDUCTION THROUGH PARTIALLY INSULATED CEILINGS

Area of Ceiling
(Take area of upstairs ceiling in a two-story house)

Ceiling area will normally be the same as floor area (from building description sheet)

If ceilings are partially insulated, e.g. to no more than R-10, further savings can be made by adding insulation. However the primary savings have already been made and the investment in extra insulation may not be as worthwhile as some other measures. Evaluate the savings by the calculations below, using R-10 for a sample building or actual R value for a sample building or actual R value for an individual case.

Probable Heat Loss through Partially Insulated Ceilings

| Ceiling area sq. ft. | × | District heating factor | + | Total R-value (Use R-10 for sample houses) | = | Heating units required |

Potential Savings by Additional Insulation of Ceilings

The appropriate amount of insulation for a ceiling depends on climate, fuel prices and insulation costs. Check the current DOE recommendations on maximum appropriate R value for your area. This indicates the maximum amount of insulation which is cost effective for your climate at current prices. Any less amount of insulation than this will save less total fuel but will save more per dollar spent.

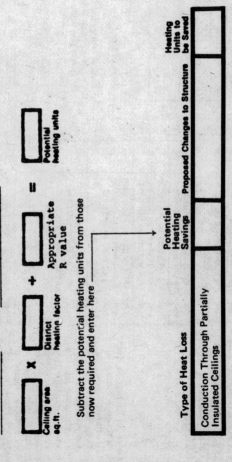

Ceiling area sq.ft. × District heating factor + Appropriate R value = Potential heating units

Subtract the potential heating units from those now required and enter here

Type of Heat Loss	Potential Heating Savings	Proposed Changes to Structure	Heating Units to be Saved
Conduction Through Partially Insulated Ceilings			

HEAT LOSSES BY CONDUCTION THROUGH FLOORS

Potential Savings on Heat Loss through Floors

1. Reducing below floor drafts

Unless vigorous under-floor ventilation is needed to remove moisture rising from the ground, a floor should be protected from drafts, so that it has a floor exposure factor of only 0.5. If the floor in this building has an unexposure factor over 0.5, cutting out drafts below the floor could reduce the heat loss to: →

Floor Exposure Factor

Select the appropriate factor from the description below:

Description	Factor
Building on posts or pillars with no perimeter protection	1.0
Building on posts or rocks with some protection against wind blowing under building	.8
Tight skirt or foundation wall around perimeter and less than two feet of wall or skirt exposed above grade	.5
Tight foundation wall with equivalent of R4 insulation on inside or outside of wall	.3

Material	Thickness (inches)	R Value
Interior surface	-	0.68
Interior surface	-	.68
	Total R value	

R value of floor

A typical uninsulated, uncarpeted wood floor will have an R value of about 3. Carpet and pad will increase the R value to 5.

$$\boxed{} \times \boxed{} \times \boxed{} \div \boxed{} = \boxed{}$$

Floor area (from building description) sq. ft. × Floor exposure factor × District heating factor ÷ Floor R value from above = Heating Units Required

$$\boxed{} \times 0.5 \times \boxed{} \div \boxed{} = \boxed{}$$

Floor area from above sq. ft. × 0.5 Floor exposure factor × District heating factor ÷ Floor R value from above = Potential heating units

2. Adding Insulation

EITHER (a) The perimeter wall of the crawl space or basement can be insulated (with R-4 or better) down to grade level or just below, to cut both drafts and conduction losses, reducing the floor exposure factor to 0.3. The heat loss with an insulated perimeter would be approximately:

Floor area from above sq. ft.	×	0.3 Floor exposure factor	×	District heating factor	÷	Floor R value from above	=	Potential heating units

OR (b) Where perimeter walls cannot be insulated because the insulation would become wet (if water seeps through the basement walls or if water rises above the level of the insulation or if underfloor ventilation is needed) the whole floor may be insulated instead of the perimeter. This is usually more expensive than perimeter insulation but can give extra savings. Insulating this floor with a minimum of 3½" fiberglass between the joists will increase the R value to 15. More insulation will increase the R value further still. The potential heating units will be:

Floor area from above sq. ft.	×	Floor exposure factor (Use 0.5 if drafts can be eliminated. Use .8 if underfloor ventilation is needed)	×	District heating factor	÷	Floor R value (minimum of 15)	=	Potential heating units with whole floor insulation

Do not complete this page for concrete slab floors. Heat loss due to them is usually small but cannot be evaluated by this method.

FILL IN AT JOB SITE.

Subtract the potential heating units for any of the three possible floor treatments which are feasible from the heating units now required and enter in the boxes below.

1	2a	2b
Potential heating savings by reducing drafts	Potential heating savings by Perimeter Insulation	Potential heating savings by whole floor insulation

HEAT LOSSES BY CONDUCTION THROUGH UNINSULATED WALLS

R Value of Outside Walls

List below all materials in walls, starting from inside and including air spaces within the wall. Insert R value for each component from Table 1.

R Value of Outside Walls

Uninsulated frame walls usually have a total R value around 3.0 no matter what type of interior finish or siding treatment is present. If the wall is of different construction, list all materials in the wall the table at right, starting from inside and including air spaces within the wall. Insert R value for each component from Table 1.

Material	Thickness (inches)	R Value
Interior surface	–	0.68
Outside surface	–	.17
Total R value		

FILL IN AT JOB SITE

[ ] Total perimeter of outside wall ft. × [ ] Total height of outside wall ft. = [ ] Gross wall area sq.ft

[ ] Gross wall area sq.ft – [ ] Total area of all windows and doors = [ ] Net wall area sq.ft.

(Partially insulated frame walls cannot normally be further insulated.)

☐ × ☐ ÷ ☐ = ☐
Net wall area District heating factor Total R Value Heating units required

Potential Savings by Insulation

Well-insulated walls should have an R value of 15. If this were so for this building, the wall heat loss would be:

☐ × ☐ ÷ **15** = ☐
Net wall area (from box above) District heating factor R value Potential heating units

Subtract the potential heating units from those now required and enter here →

Type of Heat Loss	Heating Units Required	Potential Heating Savings	Proposed Changes to Structure	Heating Units to be Saved
Conduction Through Uninsulated Walls				

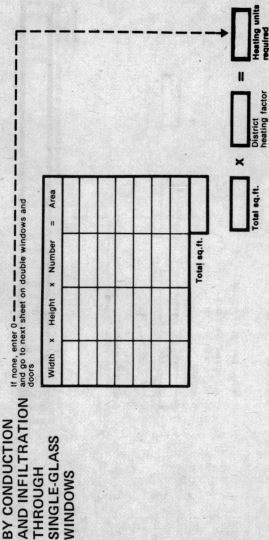

**HEAT LOSSES
BY CONDUCTION
AND INFILTRATION
THROUGH
SINGLE-GLASS
WINDOWS**

Area of Single Glass Windows:
(Assuming R = 1 for single glass)

If none, enter 0 — — — — — —
and go to next sheet on double windows and
doors

Width	×	Height	×	Number	=	Area
					Total sq.ft.	

Total sq.ft. × District heating factor = Heating units required

Potential Saving by Adding Storm Windows

Adding storm windows will cut the conduction heat loss by half and will probably save half again by reducing infiltration, i.e. cut heat loss to 25%. Calculate the potential heating units below.

[ ] × .25 = [ ] Potential Heating Units

Heating units now required with single glass

Potential Saving by Double Glazing

Double glazing or adding storm windows will cut the heat loss by half, so divide heating units by two, and enter here →

Type of Heat Loss	Heating Units Required	Potential Heating Savings	Proposed Changes to Structure	Heating Units to be Saved
Conduction Through Single-Glass Windows				

HEAT LOSSES BY CONDUCTION THROUGH DOUBLE-GLASS OR PLASTIC-COVERED WINDOWS AND THROUGH DOORS

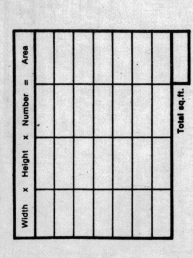

Area of Double Glass and Doors

(Assuming R = 2 for these units)

Width	×	Height	×	Number	=	Area
						Total sq.ft.

Total sq.ft. × District heating factor ÷ R value 2 = Heating units required

Potential Savings

Triple glazing of windows can be done but is not usually practical. If no change were made in the windows, the potential saving would be 0 heating units and should be entered here ⟶

(If windows were triple glazed, the R value would be approximately 3, and the potential savings would be one-third of the "Heating Units Required.")

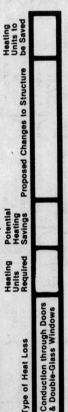

Type of Heat Loss	Heating Units Required	Potential Heating Savings	Proposed Changes to Structure	Heating Units to be Saved
Conduction through Doors & Double-Glass Windows				

The Home Energy Audit "Summary Table" for Priority Rating of Weatherization Measures requires entering the fuel cost per Heating unit in Column 4, and the Potential Heating Savings in Column 5. Enter the estimated material cost for the required changes in Column 1, the appropriate Installation factor in Column 2 and the expected life of each Weatherization measure in Column 4. From these figures determine the Total Cost and Life-time Savings from each measure, and then in Column 8 calculate the Cost:Benefit Ratio. Finally, in Column 9, rank the weatherization measures in order of cost effectiveness.

HEAT LOSS BY INFILTRATION

Building Component	One Air change per hour (1)	Two Air change per hour (2)	Three Air change per hour (3)
Cellar or	Tight, no cracks, caulked sills, sealed cellar windows, no grade entrance leaks ○	Some foundation cracks, loose cellar windows, grade entrance not tight	Major foundation cracks, poor seal around grade entrance ○
Crawl Space	Plywood floor, no trap door leaks, no leaks around water, sewer, and electrical openings	Tongue-and-groove board floor, reasonable fit on trap doors, around pipes ○	Board floor, loose fit around pipes
Windows	Storm windows with good fit ○	No storm windows, good fit on regular windows ○	No storm windows, loose fit on regular windows ○
Doors	Good fit on storm doors ○	Loose storm doors, poor fit on inside door	No storm doors, loose fit on inside door ○
Walls	Caulked windows and doors, building paper used under siding ○	Caulking in poor repair, building needs paint ○	No indication of building paper, evident cracks around door and window frame ○

Multiply the number of check marks in the first column by 1, the second column by 2, and the third column by 3. The Draft Index will be the sum of these products, divided by 4.

Record information here on Page 70 under General Heat Waste of Job Order Form

FILL IN AT JOB SITE

House Draft Index: Opposite each of the four component parts of a building in the table below, place a check mark in the circle adjacent to the features which best describe the condition of the building.

| Floor area sq.ft. | × | Height to ceiling (to upstairs ceiling in two-story house) ft. | = | Volume of air in building cu.ft. |

| Volume of air in building | × | Draft index | | | District heating factor | × .02 = | Heating units required |

Potential Savings by Reducing Infiltration

of air changes to one per hour). If the draft index for this building were improved to 1, the infiltration loss would be:

It should be possible to reduce the draft index for a building to 1 (that is, reduce the number

| Volume (from above) | × | 1 | | | District heating factor | × .02 = | Potential heating units |
| | | Draft index | | | | | |

Subtract the potential heating units from those now required and enter here

Type of Heat Loss	Heating Units Required	Potential Heating Savings	Proposed Changes to Structure	Heating Units to be Saved
Infiltration				

Summary Table

Source of Heat Loss	Weatherization Measure Required to Minimize Heat Loss	Unit Cost of Mat'l $/ft2	Quantity of Material (sq. ft.)	1 Total Material Cost (Unit Cost × Quantity) ($)	2 Installation Factor (Allows for Labor Cost to install)	3 $ Total Cost (Col 1 × Col 2)	4 Expected Life of Weatherization Measure (years) OR or Expected Life of Bldg. whichever is least
General Waste of Heat	See separate Job Order Sheet & pages 62-63					Usually Small	
Uninsulated Ceilings	Insulate to Local Standard of R =						
Partially Insulated Ceilings	Insulate to Local Standard of R =						
Exposed Floors	Reduce Exposure factor to 0.5						
*Uninsulated Perimeter	Insulate perimeter with R-4 or better						
*Uninsulated floors	Insulate floor with R-11 or better						
Uninsulated Walls	Insulate Wall to R = 15						
Single Glass Windows	Add Glass Storm Windows						

*Only one of these alternate measures is applied to a specific building. See weatherization manual for details.

Summary Table, continued

Source of Heat Loss	Weatherization Measure Required to Minimize Heat Loss	5 Potential Heat Savings from bottom of pgs. 50-63 (Heating Units/ year)	6 Fuel Cost per Heating Unit (from pg.48) ($/HU)	7 Lifetime Savings due to Weatherization Measure (Col 4 x Col 5 x Col 6) ($)	8 Cost: Benefit Ratio (Col 3 ÷ Col 7) ($)	9 Order of Cost Effectiveness
General Waste of Heat	See separate Job Order Sheet & pages 62-63	Usually Substantial			Usually Favorable	1
Uninsulated Ceilings	Insulate to Local Standard of R =					
Partially Insulated Ceilings	Insulate to Local Standard of R =					
Exposed Floors	Reduce Exposure factor to 0.5					
*Uninsulated Perimeter	Insulate perimeter with R-4 or better					
*Uninsulated floors	Insulate floor with R-11 or better					
Uninsulated Walls	Insulate Wall to R - 15					
Single Glass Windows	Add Glass Storm Windows					

*Only one of these alternate measures is applied to a specific building. See weatherization manual for details.

RECOMMENDED WEATHERIZATION PRIORITY LIST

for _____ in _____ _____

(building/type)   area/address   by Job Book Reference No. _____

PRIORITY	SOURCE OF HEAT LOSS	WEATHERIZATION MEASURE REQUIRED	Approx. Allowable Unit Cost of Materials	If list is for a Typical Building check if item is included on Job Order Sheet or note reason for omission
1	General Heat Waste	See Job Sheet	- - -	
2				
3				
4				
5				
6				
7				
8				
9				

RECOMMENDED WEATHERIZATION PRIORITY LIST: Enter the various sources of heat loss and appropriate weatherization measures from the Summary Table above in priority order in the list at right. This is the order of importance of weatherization measures in a building of the type examined regardless of size or age. A copy of this list should be included in the local Building Check and Job Order Sheet for this type of building.

BUILDING CHECK AND JOB ORDER SHEET

Each block on this sheet provides basic order and control information for the various weatherization jobs on a building. The sheet can be used intact, or individual blocks may be pasted up to provide a "camera ready" form for duplication, which covers the specific weatherization measures agreed between the D.O.E. Regional Office and the State for a particular building type in an area. If a specific measure in a block is not to be undertaken, e.g. repairing hot water faucets, it may be omitted from the duplicated form.

During inspection of building, cross out any item which does not apply or is already weatherized. Fill in appropriate instruction for jobs to be done.

JOB _____   PHONE # _____

NAME _____

ADDRESS _____

DIRECTIONS AND SPECIAL PROBLEMS

INTAKE	/	/
ESTIMATE	/	/
APPROVED	/	/
WINDOWS ORDERED	/	/
WINDOWS RECEIVED	/	/
STARTED	/	/
COMPLETED	/	/

Total _____
Materials $ _____
Cost _____

JOB ORDERS

	Estimated Quantity	Estimated Cost ($)	Actual Quantity	Actual Cost

STORM WINDOWS

FIRST FLOOR

#	width x height	no.	cost
1.	x		$
2.	x		$
3.	x		$
4.	x		$
5.	x		$
6.	x		$
7.	x		$
8.	x		$

SECOND FLOOR

#	width x height	no.	cost
1.	x		$
2.	x		$
3.	x		$
4.	x		$
5.	x		$
6.	x		$
7.	x		$
8.	x		$

OTHER

#	width x height	no.	cost
1.	x		$
2.	x		$
3.	x		$
4.	x		$
5.	x		$
6.	x		$
7.	x		$
8.	x		$

Total Costs

FLOORS | Can FLOOR EXPOSURE be reduced ?
Skirt crawl space with _____
Insulate perimeter with _____
(does perimeter insulation need to be waterproof?)
Should FLOOR INSULATION be installed?
Insulate floor with _____

CEILINGS | What is the R VALUE of the EXISTING ceiling? _____
(if existing R is over _____ no further insulation
is to be added)
What INSULATION should be added?
Insulate with _____ inches

Joist	Ceiling Area
Spacing _____ inches	_____ × _____
	_____ × _____
	_____ × _____

Is ATTIC VENTING needed?
Install _____ vents in _____
(size)   (location)

WALLS | Should WALLS be INSULATED? Are spaces accessible?
(Remember possible moisture problems with frame walls
insulated without a vapor barrier)
Insulate wall in _____ with _____
_____ with _____
_____ with _____
(room)   (material)

Totals Carried forward →

JOB ORDERS	Estimated Quantity	Estimated Cost(s)	Actual Quantity	Actual Cost
GENERAL HEAT WASTE				
Can INFILTRATION be reduced?				
Replace broken glass in				
Reset glass in				
Replace threshold in				
Pack cracks in				
Weatherstrip windows in (not if storm windows to be installed?)				
Caulk windows in				
Weatherstrip doors in				
Caulk doors in				
Install door sweeps in				
Weatherstrip attic hatch				
Repair fireplace dampers in				
Close off fireplace in				

Do HOT-AIR DUCTS pass through cold areas? Insulate ducts leading to _____			

Can DOMESTIC HOT WATER SYSTEM be made more efficient?
 Should HOT WATER HEATER temperature be reduced?
     Turn setting down to _____ ° F
 Can ELECTRIC HOT WATER HEATER be insulated?
     Insulate with R _____			

Can CONTROL of heating system be improved?
     Turn thermostat down to _____ ° F
     (explain benefit to users)
     Install night setback thermostat _____
     Relocate thermostat to _____
 Can any UNUSED ROOMS be CLOSED off?
     (remember water pipes may freeze in closed off rooms)
     Close off heating to _____			

Totals Carried forward	→		

MECHANICAL OPTIONS

Are FURNACE EFFICIENCY MODIFICATIONS possible?

List: _____

_____

Other: _____

$ Totals

Total Est. Cost	Total Actual Cost

TYPICAL COST OF MATERIALS

Material	Type	Unit Size	Unit Cost	Local Vendor	Date
Mineral wool	16" Batt	8 ft.			
" "	Loose	Cu. ft.			
Storm Window	Glass				
"	Plastic film				
Weatherstrip					
Caulking					
Strapping					

Section 3
INSULATE YOUR HOME

You can save significantly on heating if you live practically anywhere in the U.S.A.

Do you need to adjust your thermostat?

Do you need to put on storm windows or service your oil furnace?

Do you need to weatherstrip?

CAULKING
AND WEATHERSTRIPPING

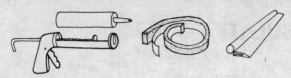

Caulking and weatherstripping are good cheap ways to save energy. It's worth your while to check to see if you need caulking, putty, or weatherstripping on your windows and doors.

DO THEY NEED CAULKING OR PUTTY?

Look at a typical window and a typical door. Look at the parts shown in the pictures. Check the box next to the description that best fits what you see:

☐ OK . . . All the cracks are completely filled with caulking. The putty around the window panes is solid and unbroken; no drafts.

☐ FAIR . . . The caulking and putty are old and cracked, or missing in places; minor drafts.

☐ POOR . . . There's no caulking at all. The putty is in poor condition; noticeable drafts.

If you checked either "FAIR" or "POOR", then you probably need caulking.

DO THEY NEED WEATHERSTRIPPING?

A. YOUR WINDOWS

Look at the parts shown in the pictures of one or two of your typical windows. Check one:

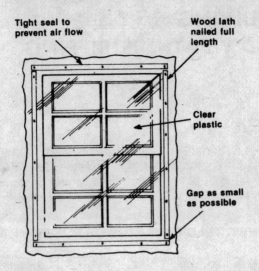

- [ ] OK ... Good, unbroken weatherstripping in all the indicated places; no drafts.
- [ ] FAIR ... Weatherstripping damaged or missing in places; minor drafts.
- [ ] POOR ... No weatherstripping at all; noticeable drafts.

If you checked either "FAIR" or "POOR", then your windows probably need weatherstripping.

Be careful, they may be in such poor condition that weatherstripping can't be installed.

B. YOUR DOORS

Look at the parts of your doors shown in the picture.

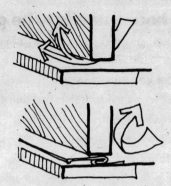

☐ OK ... Good, unbroken weatherstripping in all the indicated places; no drafts.

☐ FAIR ... Weatherstripping damaged or missing in places; minor drafts.

☐ POOR ... No weatherstripping at all; noticeable drafts.

If you checked either "FAIR" or "POOR", then your doors probably need weatherstripping.

CAULK THE OPENINGS IN YOUR HOME

Caulking should be applied wherever two different materials or parts of the house meet. It takes no specialized skill to apply and a minimum of tools.

Where a house needs to be caulked

1. Between window drip caps (tops of windows) and siding.

2. Between door drip caps and siding.

3. At joints between window frames and siding.

4. At joints between door frames and siding.

5. Between window sills and siding.

6. At corners formed by siding.

7. At sills where wood structure meets the foundation.

8. Outside water faucets, or other special breaks in the outside house surface.

9. Where pipes and wires penetrate the ceiling below an unheated attic.

10. Between porches and main body of the house.

11. Where chimney or masonry meets siding.

12. Where storm windows meet the window frame, except for drain holes at window sill.

13. And if you have a heated attic; where the wall meets the eave at the gable ends.

Materials

What you'll need

Caulking compound is available in these basic types:

1. Oil or resin base caulk; readily available and will bond to most surfaces — wood, masonry and metal; not very durable but lowest in first cost for this type of application.

2. Latex, butyl or polyvinyl based caulk; all readily available and will bond to most surfaces, more durable, but more expensive than oil or resin based caulk.

How much

Estimating the number of cartridges of caulking compound required is difficult since the number needed will vary greatly with the size of cracks to be filled. Rough estimates are:

> 1/2 cartridge per window or door
>
> 2 cartridges for a two story chimney
>
> 4 cartridges for the foundation sill

Installation

1

Before applying caulking compound, clean area of paint build-up, dirt, or deteriorated caulk with solvent and putty knife or large screwdriver.

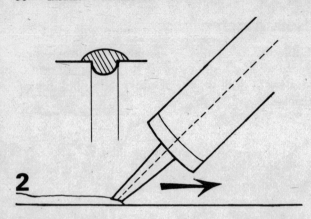

Drawing a good bead of caulk will take a little practice.
First attempts may be a bit messy. Make sure the bead
overlaps both sides for a tight seal.

A wide bead may be necessary to make sure caulk
adheres to both sides.

4

Fill extra wide cracks like those at the sills (where the house meets the foundation) with oakum, glass fiber insulation strips, etc.)

FOUNDATION SILL

5

In places where you can't quite fill the gaps, finish the job with caulk.

6

Caulking compound also comes in rope form. Unwind it and force it into cracks with your fingers. You can fill extra long cracks easily this way.

WEATHERSTRIP YOUR WINDOWS

Weatherstripping windows can be accomplished by even the inexperienced handyman. A minimum of tools and skills is required.

But before starting, make sure that both the moving parts of your windows (the sash), and the channels that the sash slide in aren't so rotted that they won't hold the small nails used for weatherstripping. If they are badly rotted, don't weatherstrip, but consider replacing the entire window unit first. Call your lumberyard or window dealer for an evaluation or cost estimate.

Materials
What you'll need

Thin spring metal

Installed in the channel of window so it is virtually invisible. Somewhat difficult to install. Very durable.

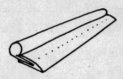

Rolled vinyl

With or without metal backing. Visible when installed. Easy to install. Durable.

Foam rubber with adhesive backing

Easy to install. Breaks down and wears rather quickly. Not as effective a sealer as metal strips or rolled vinyl.

Never use where friction occurs.

How much

Weatherstripping is purchased either by the running foot or in kit form for each window. In either case you'll have to make a list of your windows, and measure them to find the total length of weatherstripping you'll need.

Measure the total distance around the edges of the moving parts of each window type you have, and complete the list below:

Type	Size	Quantity	X	length req'd	=	Total
1. Double-hung	1	(_____)	X	(_____)	=	_____
	2	(_____)	X	(_____)	=	_____
	3	(_____)	X	(_____)	=	_____
2. Casement	1	(_____)	X	(_____)	=	_____
	2	(_____)	X	(_____)	=	_____
	3	(_____)	X	(_____)	=	_____
3. Tilting	1	(_____)	X	(_____)	=	_____
	2	(_____)	X	(_____)	=	_____
	3	(_____)	X	(_____)	=	_____
4. Sliding pane	1	(_____)	X	(_____)	=	_____
	2	(_____)	X	(_____)	=	_____
	3	(_____)	X	(_____)	=	_____

Total length of weatherstripping required _____

Be sure to allow for waste. If you buy in kit form, be sure the kit is intended for your window type and size.

PUTTY

WEATHERSTRIPPING ON INSIDE OR BASE

Installation

Thin spring metal

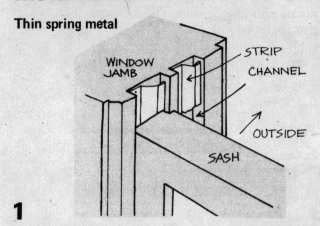

1

Install by moving sash to the open position and sliding strip in between the sash and the channel. Tack in place into the casing. Do not cover the pulleys in the upper channels.

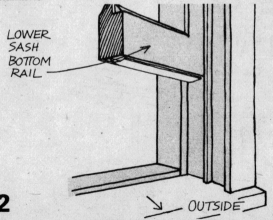

2

Install strips the full width of the sash on the bottom of the lower sash bottom rail and the top of the upper sash top rail.

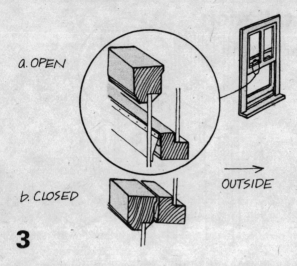

a. OPEN

OUTSIDE

b. CLOSED

3

Then attach a strip the full width of the window to the upper sash bottom rail. Countersink the nails slightly so they won't catch on the lower sash top rail.

Rolled vinyl

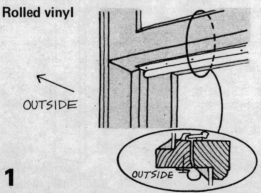

OUTSIDE

OUTSIDE

1

Nail on vinyl strips on double-hung windows as shown. A sliding window is much the same and can be treated as a double-hung window turned on its side. Casement and

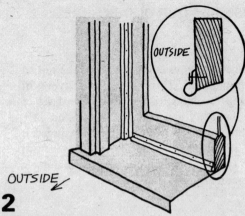

2

tilting windows should be weatherstripped with the vinyl nailed to the window casing so that, as the window shuts, it compresses the roll.

Adhesive-backed foam strip

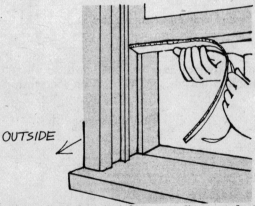

Install adhesive backed foam, on all types of windows, only where there is no friction. On double-hung windows, this is only on the bottom (as shown) and top rails. Other types of windows can use foam strips in many more places.

WEATHERSTRIP YOUR DOORS

You can weatherstrip your doors even if you're not an experienced handyman. There are several types of weatherstripping for doors, each with its own level of effectiveness, durability and degree of installation difficulty. Select among the options given the one you feel is best for you. <u>The installations are the same for the two sides and top of a door</u>, with a different, more durable one for the threshold.

1. Adhesive backed foam:

Tools

Knife or shears,
Tape measure

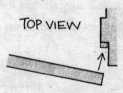

TOP VIEW

Evaluation — extremely easy to install, invisible when installed, not very durable, more effective on doors than windows.

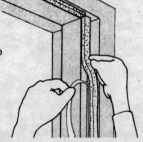

Installation — stick foam to inside face of jamb.

2. Rolled vinyl with aluminum channel backing:

Tools

Hammer, nails,
Tin snips
Tape measure

Evaluation — easy to install, visible when installed, durable.

Installation — nail strip snugly against door on the casing

3. Foam rubber with wood backing:

Tools

Hammer, nails,
Hand saw,
Tape measure

Evaluation — easy to install, visible when installed, not very durable.

Installation — nail strip snugly against the closed door. Space nails 8 to 12 inches apart.

4. Spring metal:

Tools

Tin snips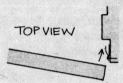
Hammer, nails,
Tape measure

Evaluation — easy to install, invisible when installed, extremely durable.

Installation — cut to length and tack in place. Lift outer edge of strip with screwdriver after tacking, for better seal.

Note: These methods are harder than 1 through 4.

5. Interlocking metal channels:

Tools

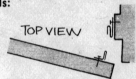

Hack saw,
Hammer, nails,
Tape measure

Evaluation — difficult to install (alignment is critical), visible when installed, durable but subject to damage, because they're exposed, excellent seal.

Installation — cut and fit strips to head of door first: male strip on door, female on head; then hinge side of door: male strip on jamb, female on door; finally lock side on door, female on jamb.

6. Fitted interlocking metal channels: (J-Strips)

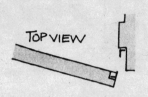

TOP VIEW

Evaluation — very difficult to install, exceptionally good weather seal, invisible when installed, not exposed to possible damage.

Installation — should be installed by a carpenter. Not appropriate for do-it-yourself installation unless done by an accomplished handyman.

7. Sweeps:

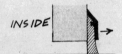

INSIDE

Tools

Screwdriver,
Hack saw,
Tape measure

Evaluation — useful for flat threshholds, may drag on carpet or rug.
Models that flip up when the door is opened are available (not illustrated).

Installation — cut sweep to fit 1/16 inch in from the edges of the door. Some sweeps are installed on the inside and some outside. Check instructions for your particular type.

8. Door Shoes:

Tools

Screwdriver,
Hack saw,
Plane,
Tape measure

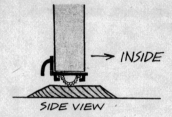

Evaluation — useful with wooden threshhold that is not worn, very durable, difficult to install (must remove door).

Installation — remove door and trim required amount off bottom. Cut to door width. Install by sliding vinyl out and fasten with screws.

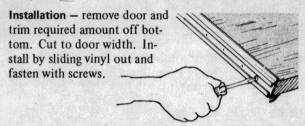

9. Vinyl bulb threshold:

Tools

Screwdriver,
Hack saw,
Plane,
Tape measure

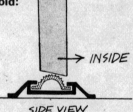

Evaluation — useful where there is no threshhold or wooden one is worn out, difficult to install, vinyl will wear but replacements are available.

Installation — remove door and trim required amount off bottom. Bottom should have about 1/8" bevel to seal against vinyl. Be sure bevel is cut in right direction for opening.

10. Interlocking threshold:

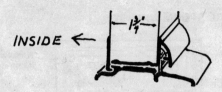

Evaluation — very difficult to install, exceptionally good weather seal.

Installation — should be installed by a skilled carpenter.

INSTALL STORM WINDOWS

There are five kinds of storm windows:

PLASTIC (POLYETHYLENE SHEET). These come in rolls and cost only 65¢ each. You may have to put up replacements each year.

Tack the plastic sheets over the outside of your windows or tape sheets over the inside instead of installing permanent type storm windows.

SINGLE PANE GLASS OR RIGID PLASTIC. These cost $25.00 for glass and $8.00 for acrylic panes. You put them up and take them down each year.

TRIPLE—TRACK GLASS (COMBINATION). These have screens and you can open and close them. They are for double-hung or sliding windows. They cost about $33.00 each installed. They are available for less without screens.

All five kinds are about equally effective. The more expensive ones are more durable, attractive, and convenient.

PLASTIC STORM WINDOWS

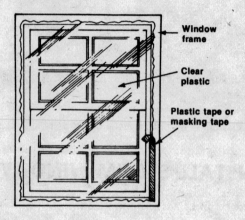

Window frame

Clear plastic

Plastic tape or masking tape

Installation

Measure the width of your larger windows to determine the width of the plastic rolls to buy. Measure the length of your windows to see how many linear feet and therefore how many rolls or the kit size you need to buy.

Attach to the inside or outside of the frame so that the plastic will block airflow that leaks around the moveable parts of the window. If you attach the plastic to the outside use the slats and tacks. If you attach it to the inside masking tape will work.

Inside installation is easier and will provide greater protection to the plastic. Outside installation is more difficult, especially on a 2 story house, and the plastic is more likely to be damaged by the elements.

Be sure to install tightly and securely, and remove all excess — besides looking better, this will make the plastic less susceptible to deterioration during the course of the winter.

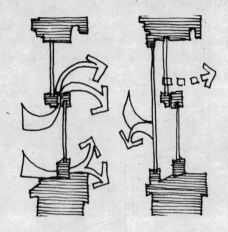

SINGLE PANE GLASS OR RIGID PLASTIC

Rigid Plastic: These are available in do-it-yourself kits.

Glass: Storm window suppliers will build single pane glass storm windows to your measurements that you then install yourself. Another method is to make your own with do-it-yourself materials available at some hardware stores.

Installation

Rigid Plastic: These are always installed on the inside. Follow the instructions on the do-it-yourself kit.

Glass: These can be installed either inside, if the way the window is built will permit it, or on the outside. If you install them on the inside, then you won't be able to open the existing window. If you install them on the outside, then they only cover the moving part of the window and you'll save less energy, but they will be permanently installed. With metal casement windows, exterior installation of single-pane storm windows is a job for a contractor.

Determine how you want the windows to sit in the frame. Your measurements will be the outside measurements of the storm window. Be as accurate as possible, then allow 1/8" along each edge for clearance. You'll be responsible for any errors in measurement, so do a good job.

When your windows are delivered, check the actual measurements carefully against your order.

Install the windows and fix in place with moveable clips so you can take them down for cleaning.

Selection: Judging Quality

Frame finish (glass windows): A mill finish (plain aluminum) will oxidize quickly and degrade appearance. Windows with an anodized or baked enamel finish look better.

Weatherstripping: The side of the storm window frame which touches the existing window frame should have a permanently installed weatherstrip or gasket to make the joint as airtight as possible.

AWNING STORM WINDOW on INSIDE

CRANK HOLE

TRIPLE—TRACK GLASS (COMBINATION).

Triple track combination (windows and screen) storm windows are designed for installation over double-hung and sliding windows. They are permanently installed and can be opened at any time for ventilation.

Double-track combination units are also available and they cost less. Both kinds are sold almost everywhere, and can be bought with or without the cost of installation.

Installation

You can save a few dollars (10% to 15% of the purchase price) by installing the windows yourself. But you'll need some tools: caulking gun, drill, and screw driver. In most cases it will be easier to have the supplier install your windows for you, although it will cost more.

The supplier will first measure all the windows where you want storm windows installed. It will take anywhere from several days to a few weeks to make up your order before the supplier returns to install them.

Installation should take less than one day, depending on how many windows are involved. Two very important items should be checked to make sure the installation is properly done.

Make sure that both the window sashes and screen sash move smoothly and seal tightly when closed after installation. Poor installation can cause misalignment.

Be sure there is a *tightly* caulked seal around the edge of the storm windows. Leaks can hurt the performance of storm windows a lot.

NOTE: Most combination units will come with two or three 1/4" dia. holes (or other types of vents) drilled through the frame where it meets the window sill. This is to keep winter condensation from collecting on the sill and causing rot. Keep these holes clear, and drill them yourself if your combination units don't already have them.

Selection: Judging Quality

Frame finish: A mill finish (plain aluminum) will oxidize, reducing ease of operation and degrading appearance. An anodized or baked enamel finish is better.

Corner joints: Quality of construction affects the strength and performance of storm windows. Corners are a good place to check construction. They should be strong and air tight. Normally overlapped corner joints are better than mitered. If you can see through the joints, they will leak air.

Sash tracks and weatherstripping: Storm windows are supposed to reduce air leakage around windows. The depth of the metal grooves (sash tracks) at the sides of the window and the weatherstripping quality makes a big difference in how well storm windows can do this. Compare several types before deciding.

Hardware quality: The quality of locks and catches has a direct effect on durability and is a good indicator of overall construction quality.

INSTALL STORM DOORS

Combination (windows and screen) storm doors are designed for installation over exterior doors. They are sold almost everywhere, with or without the cost of installation.

Installation

You can save a few dollars (10% to 15% of the purchase price) by installing doors yourself. But you'll need some tools: hammer, drill, screw driver, and weatherstripping. In most cases, it will be easier to have the supplier install your doors himself.

The supplier will first measure all the doors where you want storm doors installed. It will take anywhere from several days to a few weeks to make up your order before the supplier returns to install them. Installation should take less than one-half day.

Before the installer leaves, be sure the doors operate smoothly and close tightly. Check for cracks around the jamb and make sure the seal is as air-tight as possible. Also, remove and replace the exchangeable panels (window and screen) to make sure they fit properly and with a weather tight seal.

Selection: Judging Quality

Door finish: A mill finish (plain aluminum) will oxidize, reducing ease of operation and degrading appearance. An anodized or baked enamel finish is better.

Corner joints: Quality of construction affects the strength and effectiveness of storm doors. Corners are a good place to check construction. They should be strong and air tight. If you can see through the joints, they will leak air.

Weatherstripping: Storm doors are supposed to reduce air leakage around your doors. Weatherstripping quality makes a big difference in how well storm doors can do this. Compare several types before deciding.

Hardware quality: The quality of locks, hinges and catches should be evaluated since it can have a direct effect on durability and is a good indicator of overall construction quality.

Construction material: Storm doors of wood or steel can also be purchased within the same price range as the aluminum variety. They have the same quality differences and should be similarly evaluated. The choice between doors of similar quality but different material is primarily up to your own personal taste.

BUYING OF INSULATION
R-VALUE OF THE INSULATION

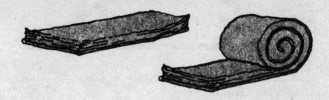

UNFINISHED ATTIC, NO FLOOR

Batts, blankets or loose fill in the floor between the joists:

THICKNESS OF EXISTING INSULATION	HOW MUCH TO ADD	HOW MUCH TO ADD IF YOU HAVE ELECTRIC HEAT **OR** IF YOU HAVE OIL HEAT AND LIVE IN A COLD CLIMATE *	HOW MUCH TO ADD IF YOU HAVE ELECTRIC HEAT AND LIVE IN A COLD CLIMATE **
0"	R-38	R-38	R-38
0"-2"	R-22	R-30	R-38
2"-4"	R-11	R-11	R-30
4"-6"	R-11	R-11	R-19
6"-8"	None	None	R-11

*Add this much if:
A. You're doing it yourself and your Heating and Cooling Factors add up to more than 0.4, or
B. You're hiring a contractor and your Heating and Cooling Factors add up to more than 0.6.

**Add this much if:
A. You're doing it yourself and your Heating and Cooling Factors add up to more than 0.7, or
B. You're hiring a contractor and your Heating and Cooling Factors add up to more than 1.0.

FINISHED ATTIC

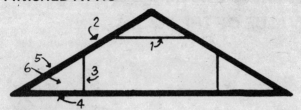

1. Attic Ceiling — see the table at the left under Unfinished Attic, No Floor.

2. Rafters — contractor fills completely with blow-in insulation.

3. Knee Walls — Insulate (5); Outer Attic Rafters instead.

4. Outer Attic Floors — Insulate (5), Outer Attic Rafters instead.

5. Outer Attic Rafters — Add batts or blankets: If there is existing insulation in (3) and (4), add R-11. If there is no existing insulation in (3) and (4), add R-19.

6. End Walls — Add batts or blankets, R-11.

UNFINISHED ATTIC WITH FLOOR

A. Do-it-yourself or Contractor Installed:

Between the collar beams — follow the guidelines above in Unfinished Attic, No Floor.

Rafters and end walls — buy insulation thick enough to fill the space available (usually R-19 for the rafters and R-11 for the end walls).

B. Contractor Installed

Contractor blows loose-fill insulation under the floor. Fill this space completely.

FRAME WALLS — contractor blows in insulation to fill the space inside the walls.

CRAWL SPACE — R-11 batts or blankets against the wall and the edge of the floor.

FLOORS — R-11 batts or blankets between the floor joists, *foil-faced.*

BASEMENT WALLS — R-11 batts or blankets between wall studs.

What kind of insulation

BATTS— glass fiber, rock wool

Where they're used to insulate:

unfinished attic floor

unfinished attic rafters

underside of floors

best suited for standard joist or rafter spacing of 16" or 24", and space between joists relatively free of obstructions

cut in sections 15" or 23" wide, 1" to 7" thick, 4' or 8' long

with or without a vapor barrier backing   — if you need one and can't get it, buy polyethylene except that to be used to insulate the underside of floors

easy to handle because of relatively small size

use will result in more waste from trimming sections than use of blankets

fire resistant, moisture resistant

FOAMED IN PLACE— ureaformaldehyde-based

Where it's used to insulate:

finished frame walls

moisture resistant, fire resistant

may have higher insulating value than blown-in materials

more expensive than blown-in materials

quality of application to date has been very inconsistent — choose a qualified contractor who will guarantee his work.

BLANKETS— glass fiber, rock wool

Where they're used to insulate:

unfinished attic floor

unfinished attic rafters

underside of floors

best suited for standard joist or rafter spacing of 16" or 24", and space between joists relatively free of obstructions

cut in sections 15" or 23" wide, 1" to 7" thick in rolls to be cut to length by the installer

with or without a vapor barrier backing

a little more difficult to handle than batts because of size

fire resistant, moisture resistant

RIGID BOARD— extruded polystyrene bead board (expanded polystyrene) urethane board, glass fiber

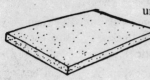

Where it's used to insulate:

basement wall

NOTE: Polystyrene and urethane rigid board insulation should only be installed by a contractor. They must be covered with 1/2" gypsum wallboard to assure fire safety.

extruded polystyrene and urethane are their own vapor barriers, bead board and glass fiber are not.

high insulating value for relatively small thicknesses, particularly urethane.

comes in 24" or 48" widths

variety of thicknesses from 3/4" to 4"

LOOSE FILL (poured-in)— glass fiber, rock wool, cellulosic fiber, vermiculite, perlite

Where it's used to insulate:

unfinished attic floor

vapor barrier bought and applied separately

best suited for non-standard or irregular joist spacing or when space between joists has many obstructions

glass fiber and rock wool are fire resistant and moisture resistant

cellulosic fiber chemically treated to be fire resistant and moisture resistant; treatment not yet proven to be heat resistant, may break down in a hot attic; check to be sure that bags indicate material meets Federal Specifications. If they do, they'll be clearly labelled.

cellulosic fiber has about 30% more insulation value than rock wool for the same installed thickness   (this can be important in walls or under attic floors).

vermiculite is significantly more expensive but can be poured into smaller areas.

vermiculite and perlite have about the same insulating value.

all are easy to install.

LOOSE FILL (blown-in)— glass fiber, rock wool, cellulosic fiber

Where it's used to insulate

unfinished attic floor

finished attic floor

finished frame walls

underside of floors

vapor barrier bought separately

same physical properties as poured-in loose fill.

Because it consists of smaller tufts, cellulosic fiber gets into small nooks and corners more consistently than rock wool or glass fiber when blown into closed spaces such as walls or joist spaces.

When any of these materials are blown into a closed space enough must be blown in to fill the whole space.

TYPE OF INSULATION

BATTS OR BLANKETS		
	glass fiber	rock wool
R-11	3½''-4''	3''
R-19	6''-6½''	5¼''
R-22	6½''	6''
R-30	9½''-10½''*	9''*
R-38	12''-13''*	10½''*

* two batts or blankets required.

LOOSE FILL (POURED-IN)			
	glass fiber	rock wool	cellulosic fiber
R-11	5''	4''	3''
R-19	8''-9''	6''-7''	5''
R-22	10''	7''-8''	6''
R-30	13''-14''	10''-11''	8''
R-38	17''-18''	13''-14''	10''-11''

INSULATE YOUR ATTIC

Attic insulation is one of the most important energy-saving home improvements you can make. This section talks about insulating 3 kinds of attics.

Unfinished Attics

Unfinished Attic without a floor. Attic isn't used at all . (This includes Attics with roof trusses in them.)

Unfinished Attic with a floor.

Finished Attics

Finished Attic that can be used for living or storage.

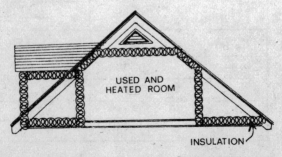

IF YOUR HOME IS A COMBINATION OF TWO KINDS OF ATTICS

(part of your attic may be finished and heated, part may be unused except as storage, as in these sample houses): If this is your situation, treat each of your attics separately.

FINISHED ATTIC

UNFINISHED, UNFLOORED ATTIC

Flat roof? Mansard roof?

If your home has a flat roof, or a mansard roof, it will be harder and more expensive to insulate than the others — talk to a contractor

INSULATE YOUR UNFINISHED UNFLOORED ATTIC

Install batts or blankets between the joists or trusses in your attic

OR

Pour in loose fill between the joists or trusses

OR

Lay in batts or pour in loose fill over existing insulation if you've decided you don't have enough already. *Don't* add a vapor barrier if you're installing additional insulation.

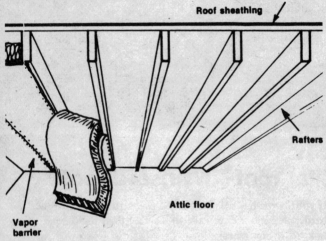

Materials
What you'll need

Batts, glass fiber or rock wool

Blankets, glass fiber or rock wool

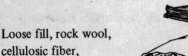

Loose fill, rock wool,
cellulosic fiber,
or vermiculite

Vapor barriers

How much

(a)   Accurately determine your attic area.

Lf necessary, divide it into rectangles and sum the areas.

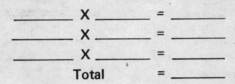

(b)   Insulation area = (.9) X (total) = _____

(c)   Vapor barrier area

1. Batts or blankets with vapor barrier backing — use insulation area.

2. Polyethylene (for use with loose fill, or if backed batts or blankets are not available) — use insulation area, but plan on waste since the polyethylene will be installed in strips between the joists or trusses, and you may not be able to cut an even number of strips out of a roll.

(d)   Insulation thickness – you may be adding two layers of insulation. Lay the first layer between the joists or trusses, and the second layer across them. *Only* the first layer should have a vapor barrier underneath it. The second layer should be an unfaced batt or blanket, loose fill, or a faced batt or blanket with the vapor barrier slashed freely.

Installation

1

Install temporary flooring and lights. Keep insulation in wrappers until you are ready to install. It comes wrapped in a compressed state and expands when the wrappers are removed.

2

Check for roof leaks, looking for water stains or marks. If you find leakage, make repairs before you insulate. Wet insulation is ineffective and can damage the structure of your home.

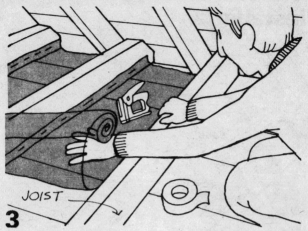

JOIST

3

Install separate vapor barrier if needed.
Lay in polyethylene strips between joists or trusses.
Staple or tack in place. Seal seams and holes with tape.
(Instead of taping, seams may be overlapped 6").

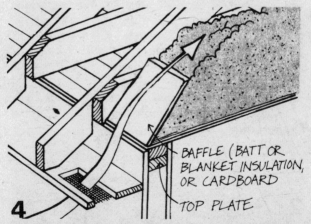

BAFFLE (BATT OR
BLANKET INSULATION,
OR CARDBOARD

TOP PLATE

4

If you're using loose fill, install baffles at the inside of
the eave vents so that the insulation won't block the
flow of air from the vents into the attic. Be sure that insulation extends out far enough to cover the top plate.

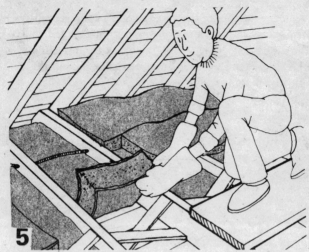

5

Lay in blankets or batts between joists or trusses. (Note: batts and blankets are slightly wider than joist spacing so they'll fit snugly). If blankets are used, cut long runs first to conserve material, using leftovers for shorter spaces. Slide insulation under wiring wherever possible.

OR

6

Pour in loose fill insulation to the depth required. If you

are covering the tops of the joists, a good way to get uniform depth is to stretch two or three strings the length of the attic at the desired height, and level the insulation to the strings. Use a board or garden rake. Fill all the nooks and crannies, but don't cover recessed light fixtures, exhaust fans, or attic ventilation.

The space between the chimney and the wood framing should be filled with *non-combustible* material, preferably unfaced batts or blankets. Also, the National Electric Code requires that insulation be kept 3" away from light fixtures.

Cut ends of batts or blankets to fit snugly around cross bracing. Cut the next batt in a similar way to allow the ends to butt tightly together.

INSULATE YOUR UNFINISHED FLOORED ATTIC

Should you insulate it? It depends on how much insulation is already there. To find out, go up there and check.

The insulation, if there is any, will be in either of two places:

Between the rafters. The first place to look is up between the rafters and in the walls at the ends of the attic.

Under the floor. If it's not up between the rafters, it might be down under the floorboards. If so, it won't be easy to see. You'll have to look around the edges of the attic, or through any large cracks in the floor. A flashlight may be handy, and also a ruler or stick that you can poke through the cracks with. If there's any soft, fluffy material in there, that's insulation.

Wherever the insulation is, if it's there at all, estimate how thick it is.

No

If it's thicker than 4 inches, it's not economical to add more.

Yes

If it's 4 inches thick or less, you might need more.

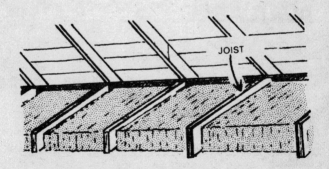

JOIST

NOTE: If you can't tell whether you have enough insulation up there, get a contractor to find out for you. You're likely to be calling one anyway to do the work, and you'll want a cost estimate from him. Ask the contractor to tell you how much insulation is already there.

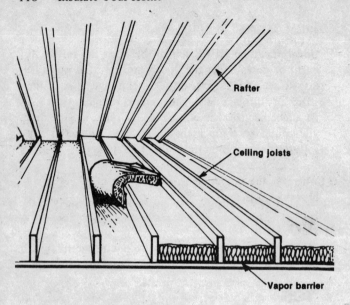

Rafter

Ceiling joists

Vapor barrier

TWO OPTIONS AVAILABLE

1. **CONTRACTOR INSTALLED:** blow-in insulation under the flooring and between the joists.

2. **DO IT YOURSELF OR CONTRACTOR:** install batts between the rafters, collar beams, and the studs on the end walls.

Types of materials contractors use

Blown-in insulation

glass fiber

rock wool

cellulosic fiber

Preparation

Do you need ventilation in your attic?

Check for roof leaks, looking for water stains or marks. If you can find any leaks, make repairs before you insulate. Wet insulation is useless and can damage the structure of your house.

What your contractor will do

The insulation is installed by blowing the insulating material under air pressure through a big flexible hose into the spaces between the attic floor and the ceiling of the rooms below. Bags of insulating material are fed into a blowing machine that mixes the insulation with air and forces it through the hose and into place. Before starting

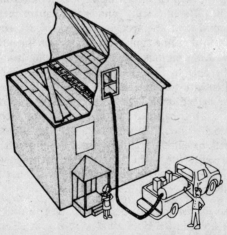

the machine, the contractor will locate the cross bracing between the joists in the attic. He'll then remove the floor boards above the cross bracing and install the insulation by blowing it in on each side of the cross bracing to make sure there are no spaces left unfilled. Since there's no effective way to partially fill a space, all of the spaces should be completely filled to ensure proper coverage. Normally the job will take no longer than a day.

What you should check

Before you sign an agreement with your contractor, decide how much and what kind of insulation you're buying and make sure it's included in the contract. Insulation material properly installed will achieve a single insulating value (R-Value) for the depth of your joist space. You should agree on what that insulating value is with the contractor, before the job begins. Next check a bag of the type of insulation he intends to use. On it, there will be a table which will indicate how many square feet of attic floor that bag is meant to cover while achieving the desired insulating value. The information may be in different forms (number of square feet per bag or number of bags per 1000 square feet), so you may have to do some simple division to use the number correctly. Knowing this and the area of your attic, you should be able to figure out how many bags must be installed to give you the desired R-Value. This number should be agreed upon between you and the contractor before the job is begun. While the job is in progress, be sure the right amount is being installed. There's nothing wrong with having the contractor save the empty bags so that you can count them (5 bags more or less than the amount you agreed on is an acceptable difference from the estimate).

After the job is finished, it's a good idea to drill 1/4" diameter holes in the floor about a foot apart. This will help prevent condensation from collecting under the floor in winter.

DO-IT-YOURSELF

Install batts or blankets in your attic between the rafters and collar beams, and the studs on the end walls.

BLOWN MINERAL
WOOL INSULATION

This measure will involve installing 2x4 beams which span between each roof rafter at ceiling height, if your attic doesn't already have them. This gives you a ventilation space above for the insulation.

NOTE: The materials, methods, and thicknesses of insulation are the same for both do-it-yourself and contractor jobs.

What you'll need

Buy either batts or blankets, made out of glass fiber or rock wool. Follow safety directions for material.

Do you need insulation with an attached vapor barrier?

Exception: For the area between the collar beams, if you're laying the new insulation on top of old insulation, buy insulation without a vapor barrier if possible, or slash the vapor barrier on the new insulation.

How Thick?

For the rafters and end walls, buy insulation that's thick enough to fill up the rafter and stud spaces. If there's some existing insulation in there, the combined thickness of the new and old insulation together should fill up the spaces.

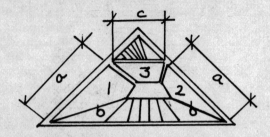

1. Figure out the area you want the insulation to cover between your rafters and collar beams (shown above). In general, figure each area to be covered, and add the areas up. If your attic is like the one shown, measure distances a, b, and c, enter them below, and do the figuring indicated (the .9 allows for the space taken up by rafters or collar beams.):

_____ x _____ x .9 = _____
distance a   distance b   Area 1

_____ x _____ x .9 = _____
distance a   distance b   Area 2

_____ x _____ x .9 = _____
distance b   distance c   **+**   Area 3

TOTAL [   ]

**total area of insulation
needed for rafters and
collar beams.**

2. Calculate the length of 2x4 stock you'll need for collar beams. Measure the length of span you need between rafters (c) and count the number of collar beams you need to install. Multiply to get the length of stock you need. You can have the lumber yard cut it to length at a small charge. If you cut it yourself, allow for waste. If you plan to finish your attic, check with your lumber yard to make sure 2" X 4" 's are strong enough to support the ceiling you plan to install.

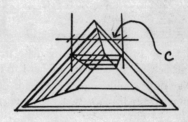

3. Figure out the area of each end wall you want to insulate. Measure (d) and (e) and multiply to determine the area. Multiply by (.9) to correct for the space taken up by the studs, then multiply by the number of end walls.

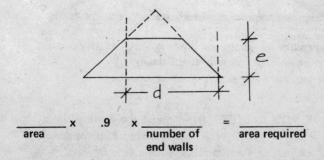

_____ x   .9   x _____   =   _____
  area   number of   area required
                        end walls

INSULATE YOUR FINISHED ATTIC

This attic is a little harder to insulate than an unfinished attic because some parts are hard to reach. A contractor can do a complete job, but if you do-it-yourself, there will probably be parts that you can't reach.

Should you insulate it?

You need to find out if there's enough insulation there already.

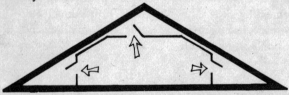

Depending on what your house is like, you may or may not be able to measure your insulation by getting into the unfinished spaces in your attic through a door or hatchway.

1. **IF YOU CAN GET IN,** measure the depth of insulation. If you have 9 inches or more of insulation everywhere, you have enough.

2. **IF YOU CAN'T GET INTO THE UNFINISHED PARTS OF YOUR ATTIC AT ALL,** have a contractor measure the insulation for you. Ask him how much is there.

When you go to take a look at these places, make a note of the depth of insulation that's already there; you'll want this information in a minute.

Which method?

You may have already found out that you can't do-it-yourself because you can't get into the unfinished part of your attic. If you can get in, there are some good things you can do yourself to insulate it.

Depending on your particular attic you may be able to do one or more of these:

A.   INSULATE ATTIC CEILING

You can insulate your attic ceiling if there's a door to the space above the finished area. You should consider insulating it if there's less than 9 inches already there.

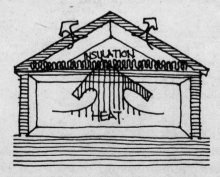

B. INSULATE OUTER ATTIC RAFTERS

"Outer attic rafters" are the parts of the roof shown in the picture below:

You should consider insulating them if:

— there's no insulation between the rafters; and

— there's room for more insulation in the outer attic floor and in the "knee walls" that separate the finished and unfinished parts of the attic.

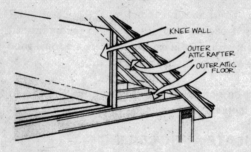

C. INSULATE OUTER ATTIC GABLES

"Outer attic gables" are the little triangular walls shown in the picture. You should insulate them if you insulate the outer attic rafters.

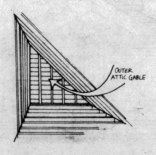

TWO OPTIONS AVAILABLE

(and worth considering if there's under 4 inches of insulation already there.)

1. **Contractor Installation:** insulation blown into the ceiling, sloping rafters and outer attic floors; batts installed in the knee walls.

2. **Do-it-yourself:** installation of batts, blankets or loose fill in all attic spaces you can get to.

Where the insulation needs to be installed

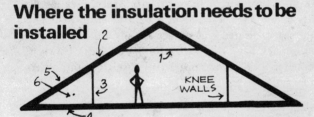

1. Attic Ceiling
2. Rafters
3. Knee Walls
4. Outer Attic Floors, or
5. Outer Attic Rafters
6. End Walls

Types of materials contractors use

Blown-in insulation
    glass fiber
    rock wool

Batts or blankets
    glass fiber
    rock wool

Preparation

How thick should the insulation be?

Check your need for ventilation and a vapor barrier.

Check for roof leaks, looking for water stains or marks. If you can find any leaks, make repairs before you insulate. Wet insulation is useless and can damage the structure of your house.

What your contractor will do

Your contractor will blow insulation into the open joist spaces above your attic ceiling, between the rafters, and into the floor of the outer attic space, then install batts in the knee walls. If you want to keep the outer attic spaces heated for storage or any other purpose, you should have the contractor install batts between the outer attic rafters instead of insulating the outer floors and knee walls.

DO-IT-YOURSELF

You can insulate wherever you can get into the unfinished spaces.

Installing insulation in your attic ceiling is the same as installing it in an unfinished attic.

If you want to insulate your outer attic spaces yourself, install batts between the rafters and the studs in the small triangular end walls.

DO YOU NEED AN ATTIC VAPOR BARRIER?

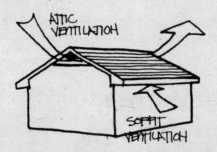

CONTRACTOR INSTALLED OR DO-IT-YOURSELF

Whenever you add insulation to your house, you should consider the need for a vapor barrier or more ventilation where you're doing the work.

A vapor barrier will prevent water vapor from condensing and collecting in your new insulation or on the beams and rafters of your house.

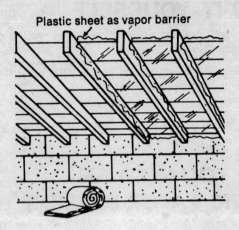

Plastic sheet as vapor barrier

Added ventilation will remove water vapor before it gets a chance to condense and will also increase summer comfort by cooling off your attic.

What you need

If you're insulating your attic and:

... you live in Zone I

1. Install a vapor barrier (unless you are blowing insulation into a finished attic)

2. Add ventilation area equal to 1/300 your attic floor area if:

   Signs of condensation occur after one heating season

   OR

   You can't install a vapor barrier with your insulation

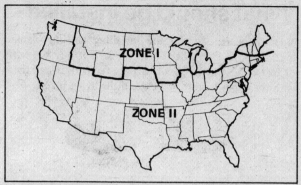

. . . you live in Zone II and have air conditioning

1. Install a vapor barrier toward the living space if you are insulating a finished attic (with other attics a vapor barrier is optional).

2. Add ventilation area equal to 1/150 your attic floor area.

. . . if you live in Zone II and don't have air conditioning

1. Install a vapor barrier toward the living space if you are insulating a finished attic (with other attics a vapor barrier is optional).

2. Add ventilation area equal to 1/300 your attic floor area if signs of condensation occur after one heating season.

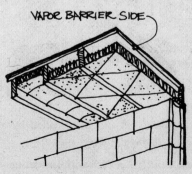

VAPOR BARRIER SIDE

What should be installed

Vapor barriers: If you are installing batt or blanket insulation, and you need a vapor barrier, buy the batts or blankets with the vapor barrier attached. Install them with the vapor barrier side toward the living space.

If you are installing a loose fill insulation, lay down polyethylene (heavy, clear plastic) in strips between the joists first.

DON'T BLOCK VENTILATION PATH

Ventilation: Install ventilation louvers (round or rectangular) in the eaves and gables (ridge vents are also available but are more difficult and costly to install in your house). The total open area of these louvers should be either 1/300 or 1/150 of your attic area and evenly divided between the gables and the eaves.

Ventilation louvers should be installed by a carpenter unless you are an experienced handyman.

Don't Block Ventilation Path with Insulation.

INSULATE WALLS

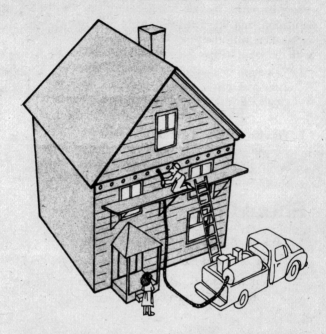

What are your walls like?

Most houses have *frame* walls. They have a wood structure — usually 2 by 4's — even though they may have brick or stone on the outside.

Some houses have brick or block *masonry* walls that form the structure of the house, without a wooden backup.

If you have frame walls, you should consider insulating them if there's no insulation at all in them already. A contractor can fill them with insulation and cut energy waste through them by 2/3.

INSULATE FLOORS

There are two cases where it's good to insulate your floor:

1. You have a crawl space that you can't seal off in winter — for example, your house stands on piers:

2. You have a garage, porch, or other cold unheated space with heated rooms above it:

Should you insulate it?

1. Is your floor uninsulated?

2. Is the floor accessible?

    · If it's above a crawl space, is the crawl space high enough for a person to work in it?

Install batts or blankets between the floor joists by stapling wire mesh or chicken wire to the bottom of the joists and sliding the batts or blankets in on top of the wire. Place vapor barrier up.

The job is quite easy to do in most cases. If you are insulating over a crawl space there may be some problems with access or working room, but careful planning can make things go much more smoothly and easily.

Check your floor joist spacing — this method will work best with standard 16" or 24" joist spacing. If you have non-standard or irregular spacing there will be more cutting and fitting and some waste of material.

How much

Determine the area to be insulated by measuring the length and width and multiplying to get the area.

(length) X (width) = area

(_____) X (_____) =_____

You may find it necessary to divide the floor into smaller areas and add them.

(length) X (width) = area

(_____) X (_____) =_____

(_____) X (_____) =_____

(_____) X (_____) =_____ +

          total area = _____

(.9)(total area) = area of insulation

(.9)(_____) =   _____

total area = area of wire mesh or chicken wire

Installation

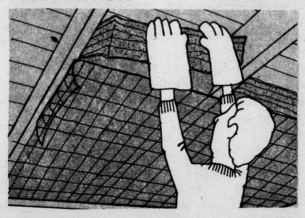

Start at a wall at one end of the joists and work out.
Staple the wire to the bottom of the joists, and at right
angles to them. Slide batts in on top of the wire. Work
with short sections of wire and batts so that it won't be
too difficult to get the insulation in place. Plan sections
to begin and end at obstructions such as cross bracing.

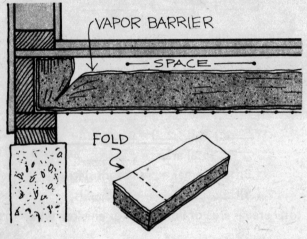

Buy insulation with a vapor barrier, and install the vapor
barrier facing up (next to the warm side) leaving an air
space between the vapor barrier and the floor. Get foil-
faced insulation if you can; it will make the air space
insulate better. Be sure that ends of batts fit snugly up
against the bottom of the floor to prevent loss of heat
up end. Don't block combustion air openings for furnaces.

RECOMMENDATIONS WORKSHEET

SPECIFICATIONS	RECOMMENDED ITEM	TO BE INCLUDED
1. Storm windows	Install_____storm windows, model_____	YES NO
2. Picture window	Install storm sash_____	YES NO
3. Storm Doors	Install_____storm doors, model_____	YES NO
4. Exterior Doors	Weatherstrip_____ exterior doors_____	YES NO
5. Ceiling Insulation	Add insulation to achieve a certified R-_____value	YES NO
6. Attic Ventilation	Add_____sf of attic ventilation, type_____	YES NO
7. Attic access	Install insulated opening, size_____sf	YES NO
8. Exterior walls	Insulate to achieve certified R-_____ value	YES NO
9. Floor insulation	Install insulation over unconditioned spaces R-_____	YES NO
10. Exterior walls	Seal all openings and cracks _____	YES NO
11. Basement windows	Install_____storm windows in basement, model_____	YES NO
12. Unconditioned spaces	Tape joints and insulate ducts _____	YES NO
13. Furnace/AC	Tune up with 1-year supply filters_____	YES NO
14. Clock thermostat	Install clock thermostat	YES NO
15. Humidifier	Install humidifier, model _____	YES NO
16. Water Heater	Replace with new efficient model	YES NO
17. AC Unit	Replace central AC unit	YES NO
18. Ceiling Fan	Install ceiling fan	YES NO
19. Furnace	Replace with proper-size unit	YES NO

MEASUREMENTS WORKSHEET SUMMARY

1. WINDOWS AND DOORS

Number of windows
requiring storms   no._____   avg._____ sf

Fit of windows   average_____   loose_____

Existing glazing   single_____double_____double_____
                            sealed   unsealed

Picture window   yes____ no____ size_____ sf

Number of doors
requiring storms   no. _____

Fit of door   avg._____   loose_____

Existing door type   wood_____wood_____metal_____
                             solid   hollow   insulated

Comments   _____

_____

_____

2. CEILING/ATTIC

Existing insulation   _____inches

Gross area   _____sf

Attic access   _____

Attic condition   pitch_____

flooring _____ %

storage_____
ventilation
existing_____

Comments   _____

_____

3. EXTERIOR WALLS

Existing insulation _____

Siding type   _____

Sheathing   _____

Gross wall area _____ sf

Glass and door
area proportion: normal_____ higher_____

4. BASEMENT/CRAWL SPACE

Gross exterior wall
area   _____

Percentage exposed
to outside   ¼_____   ½_____   ¾_____

Gross basement area _____

Gross crawl space area _____

Number of windows
requiring storms   no._____

5. HEATING/COOLING SYSTEM

Furnace   Description_____

                 Filter size_____
                 Tune-up
                 req's._____

A/C   Description_____
                 Tune up
                 req's._____

Ducts   lineal
                 feet and size_____

                 insulated: yes_____   no_____

Humidifier   yes_____   no_____

Thermostat setting   winter: day_____ night_____

                     summer: day_____ night_____

Section 4

HEATING AND
AIR CONDITIONING

TWO OPTIONS AVAILABLE

1. **Routine Servicing** — your serviceman should check all your heating and cooling equipment and do any needed maintenance once a year.

2. **Repair or Replacement** — some of your heating and cooling equipment may be so badly worn or outmoded that it will pay you to replace it now and get your money back in a few years.

A periodic checkup and maintenance of your heating and cooling equipment can reduce your fuel consumption by about 10 percent. Locating a good heating/cooling specialist and sticking with him is a good way to ensure that your equipment stays in top fuel-saving condition. Your local fuel supplier or heating/cooling system repair specialist are the people to call — you can find them in the Yellow Pages under:

Heating Contractors
Air Conditioning Equipment
Furnaces-Heating

Electric Heating
Oil Burner-Equipment and Service

Check out the people you contact with the Better Business Bureau and other homeowners in your area.

Once you're satisfied you're in touch with a reputable outfit, a *service contract* is the best arrangement to make. For an annual fee, this gets you a periodic tuneup of your heating/cooling system, and insures you against repairs of most components. A regular arrangement like this is the best one — the serviceman gets to know your system, and you're assured of regular maintenance from a company you know.

In this section, there are lists of items your serviceman should check for each type of heating or cooling system. Some items may vary from brand to brand, but *go over the list with your serviceman.* Also listed here are service items you can probably take care of yourself and save even more money. If you don't want to service your system yourself, *make sure* you add those items to your serviceman's list.

When you are faced with major repairs, inevitably the question comes up: should we fix what we've got, or buy new equipment? It's an important question but not difficult to answer if you consider the right things:

1. Get several estimates — the larger the job the more estimates. The special knowledge of the equipment dealer and installer is most needed here — they'll study your house, measure the walls and windows, and should give you *written* estimates.

2. Check to see what your fuel costs are now. See page 49 to estimate your heating bill if it's mixed in with other utilities.

3. Ask each contractor who gives you an estimate to tell you how many years he thinks it will take before the amount you save by having the new system equals what you paid for it. Remember, fuel costs are going up.

Furnace Maintenance

OIL BURNER

Every Year

Adjust and clean burner unit

Adjust fuel-to-air ratio for maximum efficiency

Check for oil leaks

Check electrical connections, especially on safety devices

Clean heating elements and surfaces

Adjust dampers and draft regulator

Change oil filters

Change air filter

Change oil burner nozzle

Check oil pump

Clean house thermostat contacts and adjust

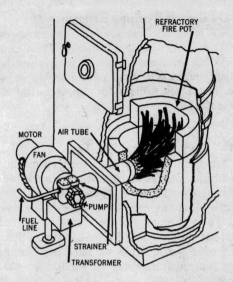

There are several tests servicemen can use to check oil furnace efficiency:

Draft Test to see if heat is being lost up the chimney or if draft is not enough to properly burn your oil.

Smoke Test to see if your oil is being burned cleanly and completely.

CO_2 test to see if fuel is being burned completely.

Stack Temperature Test to see if stack gases are too hot or not hot enough.

Is My Oil Burner Working Efficiently?

A good efficiency test involves four measurements taken with efficiency testing equipment. The old method of "eyeballing the flame" is not an accurate method of determining efficiency. Your oilman should be willing to take these measurements for you as part of his service program.

Step 1 — Stack Temperature Test

The temperature of the gases going up the chimney is an indication of whether or not you are wasting heat. Stack temperature should be between 400 to 600°F for an original oil burner and 500 to 700°F for conversion burners. Too high a temperature measured after the burner nozzle has been properly adjusted, indicates that:

The burner nozzle size is too large and more heat is being generated than can be handled by the heat exchanger; or

The heat exchanger surfaces are badly sooted. Have them brushed and vacuumed. Ask your oilman if your oil burner has a fuel oil line solenoid valve. These electrically operated valves close off the fuel supply as soon as the fire has stopped. This prevents oil from dripping into the combustion chamber causing heavy smoke and soot deposits on the heat exchanger. If you don't have one, it may pay to have one installed.

Step 2 — Smoke Test

A clean smoke test is an indication that your fuel is being burned efficiently. A dense smoke means that your fuel is not being burned completely and smoke and soot deposits are forming on the inside of your furnace. These deposits insulate the surfaces of your heat exchanger and reduce the heating efficiency of your oil burner. (A smoke reading of No. 1 is recommended, but 2 is acceptable.)

Step 3 — Draft Test

Too much draft indicates that valuable heat is being lost up the chimney. Too little draft impairs good combustion.

Most chimneys or vents produce more draft than is necessary. It is the job of the draft control to prevent this. Your oilman should measure the draft in the stack and over the fire and adjust the draft control if necessary. If you do not have a draft damper, consider installing one to improve efficiency of your system. (A stack draft of .04 and an overfire draft of .02 is recommended.)

Step 4 — CO_2 Measurement

Measurement of the carbon dioxide level in your stack gives you an indication of the combustion efficiency of the system. Oil must be thoroughly mixed with air to burn completely. Often more air is used than is actually needed to change the carbon and hydrogen in the fuel to carbon dioxide and water, which are the products of complete combustion. The amount of this excess air can be determined by measuring the amount of carbon dioxide in the stack.

Generally, the higher the carbon dioxide level, the less excess air used and the more efficient the combustion process. Too little air, however, causes smoking, increases pollution, and reduces efficiency. A carbon level of 9% is considered good. Levels over 11% are excellent. If, after a tune-up, your oilman cannot obtain a carbon dioxide reading of at least 9% without smoking, it may be that:

The furnace is leaking air into the combustion chamber and needs to be properly sealed;

There is too little or too much draft up the chimney; or

The air and oil are not thoroughly mixing for combustion.

More efficient oil burners are available now. If you have oil heating, you can check with your oil company about the new high-speed flame-retention oil burners — they can save you 10% on you oil bill.

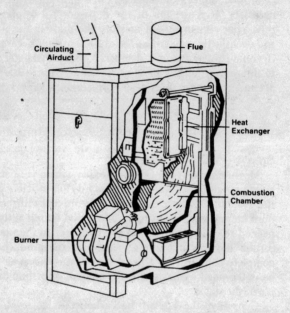

Your furnace may be too big. If your house has been insulated since it was built, then your furnace may be too big for your home. In general that means it's inefficient, and would use less fuel overall if it were smaller. Here's how to tell: wait for one of the coldest nights of the year, and set your thermostat at 70°. Once the house temperature reaches 70°, if the furnace burner runs *less* than 40 minutes out of the next hour (time it only when it's running), your furnace is too big. A furnace that's too big turns on and off much more than it should, and that wastes energy. Call your service company — depending on your type of fuel burner, they may be able to cut down the size of your burner without replacing it.

Don't overheat rooms and don't heat or cool rooms you're not using. It's important that no room in your house get more heating than it needs, and that you should be able to turn down the heating or cooling in areas of your home that you don't use. If some of your rooms get too hot before the other rooms are warm enough, you're paying for fuel you don't need, and your system needs *balancing* — call your serviceman. If your house is "zoned," you've got more than one thermostat and can turn down heating or cooling in areas where they're not needed. But if your house has only one thermostat, you can't properly adjust the temperature in rooms you're not using, and that wastes energy too. You can correct this situation fairly cheaply — try these steps on your system:

Steam Radiators — most valves on radiators are all-on or all-off, but you can buy valves that let you set any temperature you like for that radiator.

Forced-Air Heating or Cooling — Many registers (the place where the air comes out) are adjustable. If not, get ones that are, so you can balance your system.

Hot-water Radiators — if there are valves on your radiators at all, you can use them to adjust the temperature room by room.

COAL FURNACE

At the end of each heating season

Adjust and clean stoker

Clean burner of all coal, ash and clinkers

Oil the inside of the coal screw and hopper to prevent rust

GAS FURNACE (bottled, LP or natural)

Every 3 Years

Check operation of main gas valve, pressure regulator, and safety control valve

Adjust primary air supply nozzle for proper combustion

Clean thermostat contacts and adjust for proper operation

See Draft Test and Stack Temperature Test

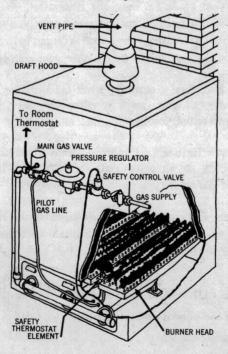

VENT PIPE →

DRAFT HOOD →

To Room
Thermostat

MAIN GAS VALVE
PRESSURE REGULATOR
SAFETY CONTROL VALVE
GAS SUPPLY

PILOT
GAS LINE

SAFETY
THERMOSTAT
ELEMENT
BURNER HEAD

ELECTRIC FURNACE

Very little maintenance required. Check the manufacturers specifications.

Heat Distribution Systems

Some items here you can do yourself to keep your system at top efficiency. For the ones you can't, check above on how to pick a serviceman. Note: except where it says otherwise, these are all once a year items.

HOT WATER HEATING SYSTEM

Serviceman:

Check pump operation

Check operation of flow control valve

Check for piping leaks

Check operation of radiator valves

Drain and Flush the boiler

Oil Pump Motor

You can do this yourself:

Bleed air from the system. Over time, a certain amount of air will creep into the pipes in your system. It will find its way to the radiators at the top of your house, and wherever there's air, it keeps out hot water. There's usually a small valve at the top of each radiator. *Once or twice a year* open the valve at each radiator. Hold a bucket under it, and keep the valve open until the water comes out. Watch out, the water is *hot*.

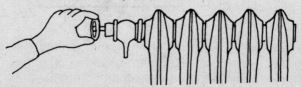

Draining and Flushing the boiler is also something you can do yourself. Ask your serviceman to show you how.

FORCED HOT AIR HEATING SYSTEM

Once a Year

Serviceman:

Check blower operation

Oil the blower motor if it doesn't have sealed bearings.

Check for duct leaks where duct is accessible.

You can do these yourself:

Clean or replace air filters — *this is important,* easy to do, and is something that needs to be done more often than it pays to have a serviceman do it. Every 30 to 60 days during the heating season you should clean or replace (depending on whether they're disposable) the air filters near the furnace in your system. Ask your serviceman how to do it, buy a supply, and stick to a schedule — you can save a lot of fuel this way:

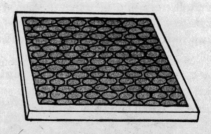

Clean the fan blade that moves the air through your system — it gets dirty easily and won't move the air well unless it's clean. Do this every year.

Keep all registers clean — Vacuum them every few weeks. Warm air coming out of the registers should have a free path unobstructed by curtains or furniture.

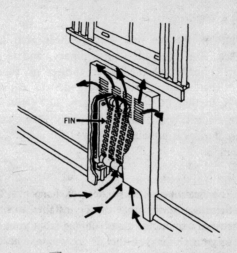

STEAM HEAT SYSTEM

With steam heat, if your serviceman checks your burner, (see Furnace Maintenance above) and the water system in your boiler, most of his work is done. There are two things you can do to save energy, though:

Insulate steam pipes that are running through spaces you don't want to heat.

Every 3 weeks during the heating season, drain a bucket of water out of your boiler (your serviceman will show you how) — this keeps sediment off the bottom of the boiler. If the sediment is allowed to stay there, it will actually *insulate* your boiler from the flame in your burner and a lot of heat will go up the chimney that would have heated your home.

New ways to get back waste heat

A lot of the fuel you buy to heat your house is wasted — it goes up your chimney, and it goes up your chimney whether your furnace is running or not. There are 2 energy-saving devices coming that can grab that heat before it gets out. (**Note:** if you have electric heat these don't apply. Also, neither of these devices is presently approved for use, but they are coming soon.)

1. **The heatpipe or stack heat exchanger** — both of these are devices that can be installed to sit in the stream of hot flue gases running from your furnace to your chimney. Either device will take heat out of the flue gas, so it can be used in the house. With warm air heating systems, the extra heat can be sent to a warm air duct. So instead of going up the chimney, the heat stays in your house.

2. **The motorized flue damper** — you know that if you leave your fireplace damper open when there's no fire going, a lot of warm air that you've paid to heat goes up the chimney — this same thing happens with your furnace when it's not running. A motorized flue damper works just like the one in your fireplace, except it's automatic — when the furnace is running, the damper's open, and the instant the furnace shuts off the damper closes.

Water Heaters

Once a Year

Serviceman:

Adjust damper (for gas or oil)

Adjust burner and clean burner surfaces (for oil)

Check electrodes (for electric)

De-lime tank

You can do this yourself:

1. Once or twice a year, drain a bucket of water out of the bottom of the heater tank — this will let out any

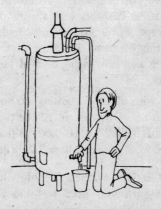

sediment that has collected there. The sediment insulates the water in the tank from the burner's flame or electrode — *that* wastes energy.

2. Insulate your water heater tank. This will greatly reduce the amount of fuel the heater uses when you
   are not using any hot water but when the heater must
   still keep the water hot. To insulate the heater, use
   3" batts or blankets with a paper or foil facing, and
   duct tape. For a more finished looking job, use duct
   insulating blankets. There are some water heater insulating kits now being sold at home improvement
   centers.

   Note: With oil or gas heaters, do *not* insulate the top
   or bottom of the heater. At the top, you may inter-
   fere with the draft of the heater's flue. At the bottom,
   you may cut off air to flame. *Only* insulate the sides.

3. Don't set the temperature of your water heater any
   higher than you need to — your heater burns fuel
   keeping your water hot when you're not using it — the
   higher you set it, the more it burns. If you've got a
   dishwasher, *140°* is high enough — if not, *120°* is plenty.
   Depending on the type of fuel you use, this simple
   setback will save you $5 to $45 a year. (You say your

heater says HIGH, MED, LOW? — Call your dealer and ask him which setting means 140 or 120 degrees.) Note: settings over 140° can shorten the life of water heaters, especially those that are glass-lined.

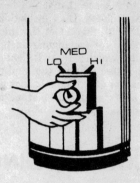

HOT WATER

All your leaky faucets should be fixed — particularly the hot ones — one leaky faucet can waste up to 6000 gallons of water a year. You can also save by turning your water heater down when you'll be away from home for a weekend or more. Always use full loads in your dishwasher and clothes washer, and use warm wash and cold rinse. Take showers — they use less hot water than baths. You should use cold water to run your garbage disposal — in general, you *save* every time you use cold water instead of hot.

Duct Insulation

If the ducts for either your heating or your air conditioning system run exposed through your attic or garage (or any other space that is not heated or cooled) they should be insulated. Duct insulation comes generally in blankets 1" or 2" thick. Get the thicker variety, particularly if you've got rectangular ducts. If you're doing this job at all, it's worth it to do it right. For air conditioning ducts, make sure you get the kind of insulation that has a vapor barrier (the vapor barrier goes on the outside). Seal the joints of the insulation tightly with tape to avoid condensation.

AIR CONDITIONING

Once a Year

(Got room air conditioners?—many of these hints apply; ask your dealer about servicing them.

Serviceman:

Oil bearings on fan and compressor if they are not sealed

Measure electrical current drawn by compressor

Check pulley belt tension

Check for refrigerating fluid leaks and add fluid if needed

Check electrical connections

Re-adjust dampers

Flush evaporator drain line.

You Can Do These Yourself

Clean or replace air filters—*this is important,* and if done every 30 to 60 days will save you far more money in fuel than the cost of the filters.

Clean the condenser coils of dust, grass clippings, etc.

Tips for Keeping Cool

Energy-saving tips for operating your air conditioner are especially important because the cost of electricity is higher in the summer months, when there is a summer surcharge on bills.

During the five-month summer billing period, May 15 to October 15, rates may be 1½¢ per kilowatt hour higher than during the seven-month winter billing period. The higher summer rates have been required because electricity demand is greater in the peak-use summer months. The purpose of summer rates is to give consumers a "price signal"—to tell them electricity does cost more to provide in the summertime, and thus to give consumers an additional incentive to conserve electricity in the summertime.

How to Choose An Air Conditioner

Two things must be considered in buying an air conditioner: the right size and the best efficiency.

An air conditioner that is too small won't keep you cool. But did you know that one that is too big for your room will turn on and off more than is necessary? That can mean higher maintenance costs and a shorter life for the unit. Also, an air conditioner that is too large won't take the moisture out of the air as well as an air conditioner of the right size. An air conditioner that is too big will cool the air quickly but not run long enough to take out the humidity.

The Right Size

Air conditioner cooling capacities are measured in Btu (British thermal units).

A typical room 9 feet wide by 12 feet long would be served by a 4,000-Btu air conditioner. A 12-by-15 room would be cooled by a 6,000-Btu unit (6,500 Btu if the room is on the sunny side.) A 15-by-18-foot room would be served by an 8,500-Btu unit (9,500 Btu, if on the sunny side.) You can choose the right size air conditioner for any room or area by using a formula. You'll need to know the dimensions of the room, or the area, to be cooled. Measure the length, height and width of the space.

Take into account insulation and exposure of the area, then use the "WHILE-divided-by-60" formula.

In the formula:

"W" stands for width in feet.

"H" is height.

"I" is insulation.

 Well-insulated = 10.

 Poorly-insulated = 18.

"L" is length.

"E" is exposure. Note the direction the longest outside wall faces. According to the direction, use these figures:

   North—16.   South—18.

   East—17.   West—20.

For example, in a room 15x8x20 feet, well insulated, with the long wall facing south, the formula would read:

W H I L E

$15 \times 8 \times 10 \times 20 \times 18 = 432,000 \div 60 = 7,200.$

So the room would need an air conditioner with a capacity of about 7,000 Btu.

Once you have figured out the size of the room air conditioner that you need, the next step is to buy one that is most efficient—that's the one that will save you money on electricity bills.

Efficiency Ratio

The Energy Efficiency Ratio (EER) is a measure of how efficiently an air conditioner uses electricity. The EER is shown as a number, usually between 7 and 12. For units of the same capacity, *the higher the EER number, the more efficient the unit.*

Stores are required by law in some states to display the EER number with air conditioners.

You can figure out the EER of any air conditioner from the data given on the unit. All you have to do is divide the number of Btu of the air conditioner model (its cooling capacity) by the number of watts listed on its name place (its electrical usage). For example, a 12,000-Btu unit using 1,500 watts would have an EER of 8.

For air conditioners of 6,000 Btu or less you should look for a unit with an EER of 7.5 or higher. For room

Unit Btu	Sample EER Range	Est. Hourly Cost	Est. Yearly Operating Cost (EYOC)
4,000	7.0	6.3¢	$ 44
	8.8	5.0¢	$ 35
5,000	7.0	7.9¢	$ 55
	8.8	6.3¢	$ 44
6,000	8.0	8.3¢	$ 58
	9.3	7.1¢	$ 50
9,000	8.0	12.4¢	$ 87
	10.7	9.3¢	$ 65
12,000	8.0	16.6¢	$116
	9.1	14.6¢	$102
15,000	8.0	20.7¢	$145
	9.3	17.9¢	$125
18,000	8.0	24.9¢	$174
	9.3	21.4¢	$150

air conditioners larger than 6,000 Btu you should buy a unit with an EER of 8.5 or more. Look for these minimum EERs, even though units with lower EERs may be sold legally.

The table above will give you an idea of the costs of running air conditioners of different EERs. You will see that the more efficient units cost less to run. The seasonal figures are based on the New York area average of

the equivalent of 700 hours of full load use during the
May 15 to October 15 period at an average cost of
11.05 cents per kilowatthour (kwh).

The labels should carry an estimated yearly operating
cost (EYOC), which will give you an approximation of
the cost of running an air conditioner during the hot
season. The labels should also carry the wattage as well
as the model number of the air conditioner.

Generally, the higher-EER, more efficient models
cost somewhat more to buy than the less efficient air
conditioners. But the savings on your electric bills will
more than make up the difference over a few seasons'
use. With the most efficient air conditioner of the
proper size, you can cut down on your summer electric
bills. As shown in the table, in one season you can save
$22−$65 as against $87−with a 9,000-Btu air condi-
tioner of 10.7 EER instead of one with an EER of 8.0.

Minimum EER Levels

Some states prohibit the sale of any new air conditioner that does not meet specified minimum EER standards. The minimum EER requirements for air conditioners will be raised each year until, by Jan. 1, 1981, no
model with an EER of less than 8 may be manufactured.

Typical Minimum EER
Permitted After Each Date

Unit Btu	Operating Voltage	Jan. 1, 1979	Jan. 1, 1980	Jan. 1, 1981
Less than 6,000	Less than 150	7.0	7.5	8.0
6,000 or more	Less than 150	8.0	8.5	8.7
Less than 24,000	150 or more	7.5	8.2	8.4
24,000 or more	150 or more	7.5	8.2	8.2

How to Use Your Air Conditioner

Air conditioners are expensive to operate. Therefore you should be careful about running them.

To cool your home as economically as possible, you need to know how to use your air conditioner . . .and when. And you need to know about other things you can do to help keep your home cool so you won't have to rely only on air conditioning.

The most important thing you can do to save electricity is to use your air conditioner only when you really need it.

Always keep your air conditioner turned off when you are away from home, or when you are not using the area that it cools. Most air conditioners will make a room comfortably cool in half an hour. An air conditioner timer, available at most hardware stores, will turn your unit on just before you come home or turn it off at any time for which you set the timer.

Set Thermostat for Air Conditioning

The next most important thing is to set the air conditioner thermostat for the highest room temperature that you find acceptable.

You should feel comfortably cool during the summer when your room temperature is about 78°. But your air conditioner thermostat is usually not marked in degrees. The control setting may be indicated by words such as "Cold," or "Colder." Take a room thermometer reading when the air conditioner is on its lowest (warmest) setting and has been running long enough to cool the area thoroughly. Put the thermometer in a central location in the room away from drafts when you take the reading. Set or reset the air conditioner's thermostat with an eye on the thermometer so that you know what setting on your unit will keep your room at 78°. Mark the setting for reference. If you are not uncomfortable at a somewhat higher temperature, by all means set the air conditioner thermostat to that.

Even if the temperature outdoors soars to an extremely high mark, it's wise to make your home no more than 15° cooler inside. That means that if it's 95° outside, you would still find relief at 80° indoors. For guidance in using your air conditioner keep a thermometer on the wall of your room and read the temperature level frequently.

How to Take Care of Your Air Conditioner

Most room air conditions have filters. The filters can be removed. Filters are located in the front, indoor side of the units. A filter may be covered by a hinged section or a removable panel. Some filters fit into slots and can be lifted out of the air conditioners. Before using your air conditioner for the first time in the hot weather season, take out the filter and clean it. Dirt, lint, soot and pollen can collect on the filters and block the air flow. This makes your unit lose efficiency.

Most filters can be washed and put back in the unit. Other filters must be discarded and replaced by new ones. If you haven't changed the filter in some time, accumulated dirt may make it fall apart when it is removed. New filters can be purchased at most hardware stores.

During the summer, filters should be cleaned or changed once a month. The most important time to clean the filter is at the start of the air conditioning season.

The condenser coils and fins (the metal grill or spines that are on the outdoor side of the unit) can become dirty or clogged. After 18 months of operation, an air conditioner can lose from 10 to 27 percent of its efficiency because the condenser is dirty. Follow the instructions in your owner's manual for taking care of this, or any other, condition of your air conditioner.

Put your air conditioner in a central location in a wall in a room—the middle instead of a corner window. Choose a window on the shady—or north side—of your home.

Keep blinds and drapes closed on the sunny side of the house. Don't put furniture, drapes or curtains so they will block circulation of air in front of air conditioner. Unnecessary lights waste electricity, add heat to the room and make the air conditioner's job harder.

THE HEAT PUMP

A heat pump runs on electricity, and is just like an air conditioner, except it can run in reverse—it can use electricity to heat, and gets more heat out of a dollar's worth of electricity than the resistance heaters in baseboard units and electric furnaces.

How? There's heat in the air outside your home, even when the temperture's below freezing, and a heat pump can get that warmth out and into your house. When should you consider installing one?—If you presently have a central electric heating system, and live south of Pennsylvania, it may pay to install a heat pump in the system, next to the furnace. Keep your electric furnace—once the temperature drops below 20° or so, the heat pump will need help from the furnace. Installation of a heat pump large enough for most houses should cost a little under $3,000, but you're getting central air conditioning as well as a "furnace" that's about 1½ times more efficient than your electric furnace.

If you're adding a room, consider adding a heat pump—like air conditioners they come in room size units. A heat pump for a room comes with its own electric resistance coil (like a baseboard electric heater) for the times of the year when it's too cold for the heat pump itself to work well. Call your air conditioner dealer for details on both central and room-size heat pumps. If your furnace runs on gas or oil, and the prices of those fuels continue to rise faster than the price of electricity, then you'll want to consider a heat pump too.

ATTIC AND ROOF

Seal any openings between your attic and the rest of your house where air might escape, such as spaces around loosely-fitting attic stairway doors or pull-down stairways, penetrations of the ceiling for lights or a fan, and plumbing vents, pipes, or air ducts which pass into the attic — they don't seem like much, but they add up!

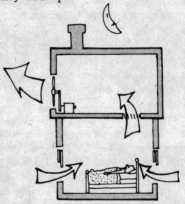

One alternative to energy-consuming air conditioning is the use of an *attic fan* to cool your home. Normally a house holds heat, so that there's a lag between the time the outside air cools after sunset on a summer night and the time that the house cools. The purpose of the attic fan is to speed up the cooling of the house by pulling air in through open windows up through the attic and out.

When the fan's on, you can let air through to the attic either by opening the attic door part way or by installing a louver that does the same thing automatically.

In the summer the sun's heat adds a large radiant heat load to the air-conditioning system. This heat will build up and be stored even through a cool summer night. Radiant heat plus stored-up heat means high summer air-conditioning bills plus strain on all components of the air-conditioning system. Adequate ventilation of attic spaces can dramatically reduce both factors and lead to both longer life for the air-conditioning system and reduced bills for electricity.

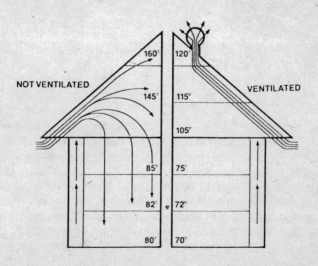

These figures represent the solar radiation intensity hitting a residence located in Kansas City, Mo. (39° N. Lat.), June 21, 1973, at 12 noon. For a 2000 ft²

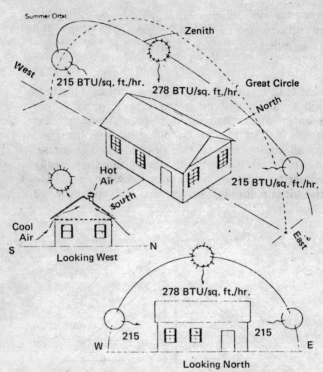

Summer Orbit

Zenith

West

215 BTU/sq. ft./hr.

278 BTU/sq. ft./hr.

Great Circle

North

215 BTU/sq. ft./hr.

East

Hot Air

South

Cool Air

S   N

Looking West

278 BTU/sq. ft./hr.

215   215

W   E

Looking North

house with a solar radiation intensity at noon of 278
Btu/ft²/h there is a heat gain of 2000 ft² times 278
Btu/ft²/h equalling 556,000 Btu/h. One ton of air
conditioning equals 12,000 Btu/h. If all the radiant
heat from the sun were to be absorbed by the house,
it would take 556,000 Btu/h divided by 12,000 Btu/h
equalling 46 tons of refrigeration to keep the house at
70° F. Why does it not take 46 tons? Part of the heat
is reflected. Part is removed by convection. Part of
the heat is stopped by insulation at the ceiling line.
The reason attic ventilators do such a dramatic job is
because there is so much heat left between the roof
and the ceiling available to be dissipated by air
removal.

For maximum flow induction, a ventilator should be located on that part of the roof where it will receive the full wind without interference. If roof ventilators are installed within the suction region created by the wind passing over the roof, their performance will be greatly increased.

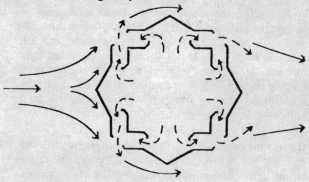

Types of roof ventilators

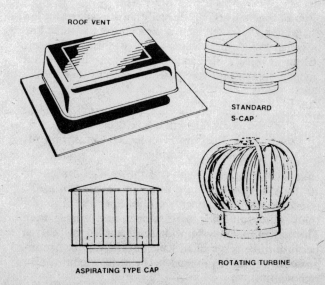

ROOF VENT

STANDARD
S-CAP

ASPIRATING TYPE CAP

ROTATING TURBINE

Hot weather energy savers

Some special summer, or warm climate saving tips:
Set air-conditioning thermostats no lower than 78°.
The 78° temperature is judged to be reasonably comfortable and energy efficient. One authority estimates that if this setting raises the temperature 6° (78° vs 72°) home cooling costs should drop about 47%. (The Federal Government is enforcing a strict 78-80° temperature in all its buildings during the summer.)

If everyone raised cooling thermostats 6° during the summer, the Nation would save more than the equivalent of 36 billion kilowatt hours of electricity, or 2% of the Nation's total electricity consumption for a year.

Run air conditioners only on really hot days and set the fan speed at high. In very humid weather, set the fan at low speed to provide less cooling but more moisture removal. Clean or replace air conditioner filters at least once a month. Turning the fan requires more electricity when the filter is dirty.

If you can confine your living spaces to fewer rooms, close off the rooms that will not be occupied.
If rooms are to be unoccupied for several hours, turn off the air-conditioning temporarily.

Buy the cooling equipment with the smallest capacity to do the job. More cooling power than necessary is inefficient and expensive. Energy-efficiency ratios (EERs) for most air-conditioning units should be available from dealers, and some window units are labeled to show the EER (the higher the EER, the more efficient the air conditioner). If you don't see a label in the showroom, ask for the information.

Cold weather energy savers

To save on heating energy and heating costs:

Lower thermostats to 68° during the day and 60° at night. If these settings reduce the temperature an average of 6°, heating costs should run about 15% less.

If every household in the United States lowered heating temperatures 6°, the demand for fuel would drop by more than 570,000 barrels of oil per day (enough to heat over 9 million homes during the winter season).

Setting nighttime temperatures back can reduce heating costs significantly. Consider the advantages of a clock thermostat which will automatically turn the heat down at a regular hour before you retire and turn it up just before you wake.

Have your furnace serviced once a year, preferably each fall. Adjustment could mean a saving of 10% in family fuel consumption.

When buying a new furnace, select one that incorporates an automatic flue gas damper, a device which reduces loss of heat when the furnace is not in operation.

If you use electric heating, consider a ''heat pump'' system. The heat pump uses outside air in both heating and cooling and can cut the use of electricity for heating by 60% or more.

Clean or replace the filter in forced-air heating systems every month.

Dust or vacuum radiator surfaces frequently.

Keep draperies and shades open in sunny windows; close them at night.

For comfort in cooler indoor temperatures use the best insulation of all—warm clothing.

Section 5

MORE ON SAVING ENERGY

Yesterday's Home

In yesterday's home energy was visible. The earliest homes were heated with wood and lit by candles and oil lamps.

By looking at the size of the wood pile the homeowner could see the amount of energy available. The log burning showed energy being used. The source of energy used for heating, cooking, and lighting was visible.

In yesterday's home the homeowner was the worker as well as the natural conserver. He brought in as much wood as he needed, and he seldom wasted it since he would have to chop more if he did. The lamp had to be refilled when it was empty so he was careful to turn it off when it was not needed. He was part of the energy cycle. More important, the cycle was one that he could visualize.

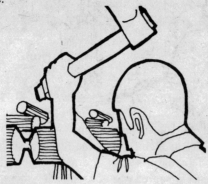

Today's Home

Let's look at the mechanical revolutions that have occurred in housing in the last 70 years. First came central heating fired by coal furnaces with hot water and steam radiators.

Then came natural gas piped into the house to fire stoves and hot water heaters as well as central heating systems.

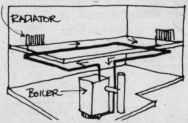

Finally, the development of electricity as an energy source led us to today's use of an ever-increasing range of appliances from air-conditioners to electric toothbrushes.

Even with the coal furnace, energy was still visible and

the homeowner was still a participant.

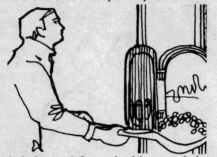

He shoveled the coal from the bin, saw it burn, and banked the fire at night. When the furnace was converted to oil or natural gas, a drastic change took place. The energy source was not visible anymore, nor was the homeowner needed in the energy cycle. More importantly, he was no longer the natural conserver. All he had to do was to pay the bills, and they seemed very modest.

Today with the cost of fuels increasing and the supply of those fuels decreasing, the homeowner must once again become the conserver. By building or retrofitting his home to conserve energy, and by using and maintaining the mechanical systems in it efficiently, he can keep his energy costs down.

THE PERSONAL ENERGY AUDIT

We have already looked at heat loss in our homes and steps in home weatherization. The Personal Energy Audit reviews our use of energy systems which produce heat, light, and monthly utility bills. This energy group includes: (1) the heating system, (2) the cooling or air conditioning system, (3) the hot-water system (for washing clothes, dishes, and people), (4) energy system for cooking and the kitchen, lighting, (5) entertainment system (television, radio, etc.), and (6) miscellaneous units (sump pump, garage door opener and other items).

	Monthly	Annual
I. HOT WATER USAGE	_____	_____

I. HOT WATER USAGE
   1. Number of baths times 30 gallons
       = _____
   2. Number of showers times minutes per
       shower times 10 gallons =_____
   3. Number of laundry loads times
       35 gallons   _____
   4. Number of dishwashing loads times
       14 gallons   _____
   5. Total gallons of heated water per week
       _____ Times 4-1/3 weeks
       _____ monthly usage

II. HEATING COSTS   _____   _____
    Analyze wood, oil, gas or electric usage

III. AIR CONDITIONING COSTS   _____   _____
     Number of hours   _____

IV. COOKING COSTS   _____   _____
    1. Number of hot meals per week _____
    2. Number and size of refrigerators_____
    3. Number and size of freezers   _____
    4. Number of appliances   _____

V. ELECTRIC COSTS   _____   _____
   Number of light hours _____

VI. ENTERTAINMENT COSTS   _____   _____
    Number of hours, TV, Radio _____

VII. OTHER COSTS   _____   _____
     Swimming pool heater   _____
     Garage Door Opener   _____
     Sump Pump   _____

YOUR ELECTRIC METER

With the precision of a finely tuned watch, your electric meter measures the amount of electricity you use in kilowatthours.

To measure your kilowatthours, your meter needs to know two things; the amount of current flowing into your house, and the pressure at which the current is flowing. Electric current flows through the wires much like water flows through a pipe; the amount of current is measured in amperes. The pressure, measured in volts, is the force that pushes the electricity through the wire. The amount of current multiplied by the voltage equals the total watts used. The number of watts used in an hour equals the total watthours. Your meter records kilowatthours (1000 watthours).

Every month, a man from the electric company comes to read your meter. Several days later, you will receives a bill for the kilowatthours you have used over the previous month. In the top section of your bill, you will find the date for your next reading.

If your meter is inside your house and you are not home on your scheduled meter reading day, the meter reader will leave a card which tells you how to read your meter and gives a phone number to call with the reading you took. This is a toll-free number. If the operator asks for your phone number, it is simply to find out where the call is coming from. By following the directions on this card, you will avoid receiving an estimated bill. However, if you don't call in the reading, your bill will be an estimate based upon your previous usage, and it will be marked "estimate."

The next time your meter is read, your correct usage will be recorded and your bill will be adjusted as necessary. Usually, estimated bills are very close to actual readings.

How to read the dials of your meter

Your meter will have either four or five dials on it. The meter most commonly used is the four dial version. At first glance, these meters look very complicated; but, after reading the instruction you should have no difficulty.

On a four-dial meter the first and third dials move counter-clockwise; the second and fourth move clockwise. The pointers move around each dial, and the reading for each one is the last number which was passed. For example, if the pointer is between five and six (regardless of which way the pointer turns), the reading is five.

Let's assume that in January, your meter looked like this: (read in the direction of the arrows)

The reading was 6125.

Now, in February, your meter had changed to the following position:

The reading was 6937.

So, your kilowatthour consumption for the February reading is the difference between the two readings (6937 - 6125) or 812 KWH. Your bill will be based on this number of kilowatthours.

The procedure is basically the same if your meter has five dials on it. The first, third and fifth pointers move clockwise, and the second and fourth move counter-clockwise. This meter is the same as a four dial meter with another dial added to the left.

If your meter looked like this in January:

The reading was 86125.

A month later, your meter had moved to this position:

The reading was 86937.

Again, your kilowatthour consumption for the one-month period is the difference between the two readings (86937 - 86125) or 812 KWH.

If you have any further questions on your electric meter, please call your local customer service center.

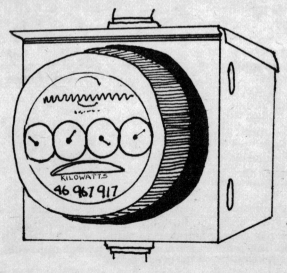

YOUR ENERGY COSTS

Obviously, a personal energy audit is a complex and highly variable undertaking.

Finding the cost of electric use begins with a review of last year's utility bills. Your old utility bills show how much electricity (or gas) you have used during the monthly period, and they indicate the cost of electricity or gas costs per unit.

1978 ANNUAL ELECTRIC COSTS†

DATE	KWH USED	CODE$	$	TOTAL	FUEL CGE.
January 9 to	521	QR	19.71		
February 8	3827	DSH	179.15	198.86	1.34
February 8 to	532	QR	19.85		
March 8	3248	DSH	152.09	171.94	1.30
March 8 to	417	QR	16.47		
April 7	2250	DSH	109.88	126.35	1.38
April 7 to	451	QR	16.88		
May 8	1201	DSH	60.85	77.73	1.22
May 8 to	37?	QR	14.16		
June 7	448	DSH	27.56	41.72	1.13
June 7 to	327	QR	12.38		
July 10	331	DSH	21.27	33.65	1.02
July 10 to	396	QR	14.71		
August 8	435	DSH	26.73	41.44	1.09
August 8 to	404	QR	16.33		
September 8	810	DSH	45.90	62.23	1.43
September 8 to	435	QR	17.76		
October 6	529	DSH	33.50	51.26	1.52
October 6 to	510	QR	19.00		
November 7	1473	DSH	74.50	93.50	1.31
November 7 to	530	QR	20.27		
December 7	2429	DSH	118.59	138.86	1.37
December 7 to	766	QR	28.02		
January 9	3632	DSH	172.60	200.62	1.36
MONTHLY	471	QR	17.96	103.18	1.29
AVERAGE	1718	DSH	85.22		

Code: QR   = Uncontrolled Water Heating (Quick Recovery)
          DSH = Residential and Space Heating
† Actual 1978 costs for a retired couple living in New England without supplemental wood heat or solar energy.

ESTIMATED USAGE

We have estimated the monthly kilowatthours used by most electric appliances, based on average usage. Because these estimates are based on a family of four, please adjust if your family is larger or smaller. Also, the use of items such as an air conditioner, electric blankets and lighting will vary with the season. As a result, you will see fluctuations in your bill seasonally.

Please keep in mind that these are *average* estimates; every customer has a somewhat different life style—affecting his use of electricity.

Because most people follow monthly budgets, we have shown usage on a monthly basis.

To determine cost averages, multiply average use by your average hourly rate.

The annual cost, for example, to operate a self-cleaning oven is 100.25 KWH x 12 months x .06 = $72.18.

Item	Estimated KWH Usage in 1 Month
FOOD PREPARATION	
Blender	1.25
Broiler	8.35
Carving Knife	.65
Coffee Maker	8.85
Deep Fryer	6.9
Dishwasher	30.25
Frying Pan	15.5
Hot Plate	7.5
Mixer	1.1
Range w/oven	97.8
with self-cleaning oven	100.25
Roaster	17.1
Toaster	2.75
Waste Disposal	2.5

FOOD PRESERVATION

15 cu. ft. Freezer	99.6
(frostless)	146.75
12 cu. ft. Refrigerator	60.65
(frostless)	101.4
14 cu. ft. Refrigerator/	
Freezer	94.75
(frostless)	152.4

LAUNDRY

Clothes Dryer	82.75
Iron (hand)	12.0
Washing Machine (automatic)	8.55
Water Heater	415.0

COMFORT & HEALTH

Air Conditioner (room)	115.75
Blanket	15.0
Dehumidifier	31.4
Fan (attic)	24.25
Fan (circulating)	3.6
Fan (window)	14.15
Hair Dryer	1.15
Humidifier	13.6
Shaver	.15
Lights (Equivalent of five 100 watt bulbs burning 5 hrs./day)	75.0

HOME ENTERTAINMENT

Radio	7.15
Radio/phono	9.1
Television (b & w)	30.15
Television (color)	41.85

HOUSEWARES

Clock	1.4
Sewing Machine	.9
Vacuum Cleaner	3.85

ELECTRICITY

If everyone scheduled household chores so as to lighten the load at the generating plants during peak hours, fewer inefficient generating units would have to be placed in service, and the utilities' daily fuel consumption would be reduced. So would the possibilities of brownouts and blackouts.

Plan your lighting sensibly. Reduce lighting were possible, concentrating it in work areas or reading areas where it is really needed. Fluorescent bulbs should be used rather than the incandescent kind.

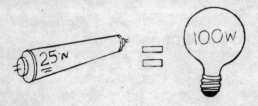

A 25-watt fluorescent bulb gives off as much light as a 100-watt incandescent bulb, but costs one fourth as much to light. Decorative gas lanterns should be turned off or converted to electric lamps. They will use much less energy to produce the same amount of light.

Careful use of lighting provides the homemaker other conservation opportunities. To save electricity through wise lighting:

Remove one bulb out of three and replace it with a burned-out bulb for safety; replace others with bulbs of the next lower wattage. But be sure to provide adequate lighting for safety (e.g., in stairwells). Concentrate light in reading and working areas, and for safety.

This should save about 4% in electricity costs in the average home.

Turn off all lights when not needed. [One 100-watt bulb burning for 10 hours uses 11,600 Btu's, or the equivalent of a pound of coal or one-half pint of oil.]

Use fluorescent lights in suitable areas—on the desk, in the kitchen and bath, among others. They give more lumens per watt. One 40-watt fluorescent tube, for example, provides more light than three 60-watt incandescent bulbs. (A 40-watt fluorescent lamp gives off about 80 lumens per watt; a 60-watt incandescent gives off only 14.7 lumens per watt. The lower-watt but higher-lumen fluorescent would save about 140 watts of electricity over a period of 7 hours.)

POSTMASTER: RETURN POSTAGE GUARANTEED

MAKE PAYMENT TO

AJAX

06591

P.O. BOX 1234
HOMETOWN, USA   77001

PUBLIC, JOHN Q. 228 EMERSON				1
FEB19	6876	66	C	17.92

5 5 30 41562394043   1792

AFTER MAR 11 76 PAY   1971
    RES 500 0 1

MAINTENANCE OF MECHANICAL SYSTEMS

What does this retrofit item do?

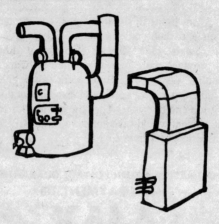

Heating and air-conditioning systems must be kept clean and properly adjusted for efficient energy use. One-quarter inch of soot on a boiler can reduce its efficiency 25%. Insulating heat pipes and ducts prevents heat loss. Fuel can be saved by lowering the nighttime temperature automatically with a clock thermostat.

What are the savings?

Lowering the nighttime room temperature 5 degrees with the use of a clock thermostat can save 10% to 12% of the fuel bill.

Energy Inspection

Check out the heating and cooling systems. If you are qualified, you may, with the homeowner's approval, clean and adjust the heating and cooling systems, replace filters, and so forth. Make a note of any uninsulated ducts or pipes running through unheated or uncooled areas of the house. Measure them for insulation.

Technical Data

Carbon build-up on forced air heaters lowers efficiency. It must be cleaned off.

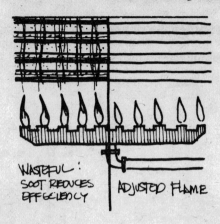

WASTEFUL:
SOOT REDUCES
EFFICIENCY   ADJUSTED FLAME

Nozzles on oil-fired furnaces will enlarge with wear and should be replaced yearly.

Air-conditioner filters should be changed or cleaned regularly.

Thermostat Setting Devices

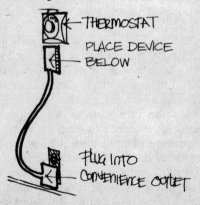

THERMOSTAT

PLACE DEVICE
BELOW

PLUG INTO
CONVENIENCE OUTLET

One type of setback device is placed below the regular thermostat. It works on a timer and heats the thermostat just enough to prevent it from turning on the heat. This produces the same results as turning the thermostat down at night.

This device can be plugged into an existing convenience outlet. It can only be used if there is an existing thermostat on the heating system.

Clock Thermostat

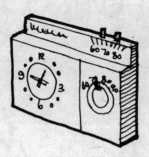

A clock thermostat installation requires an electric connection to the boiler or furnace.

WATER HEATER

Now you can buy for your home water heater an insulation refit kit that can save both money and fuel. The kit consists of a blanket of fiberglass insulation and comes with do-it-yourself installation instructions. The additional insulation on the water heater reduces the rate of heat loss, which also reduces the energy required to keep the water at the desired temperature. Studies show that, on a national average, the use of the kit will save $5 to $20 per year on your utility bill and repay your initial investment (under $20) in energy savings within 15 months for electric water heaters and within 3½ years for gas water heaters. Also, a refit kit can be used safely on electric water heaters as well as on gas water heaters when installation instructions are followed carefully. *Extreme care must be exercised when installing the insulation of a gas water heater.*

Just as a blanket or a coat helps maintain your body temperature, a storage tank protected by additional layers of insulation maintains water temperature better and longer.

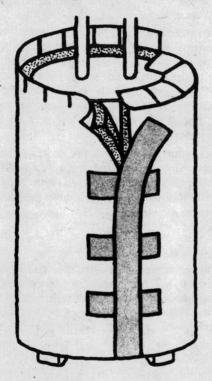

The refit kit is designed for easy installation.

How can I get a well-insulated water heater?

When buying a new water heater, you have the opportunity to choose one with a well-insulated lining. This usually means selecting the one with the thickest available insulation. However, most homes already have a water heater. Refitting it with an outside layer of fiberglass insulation will improve its energy efficiency and save money on utility bills.

How much energy could I save a year using a refit kit?

With 1½ inches of additional insulation, heat loss can be cut by about 400 kilowatt hours per year for electric water heaters and by about 3600 cubic feet per year for gas water heaters.

Note: In lieu of purchasing a refit kit, you may find it more economical to buy blanket-type insulating material to wrap the tank of an electric water heater. Of course, use of the kit is more convenient and may result in a neater appearance. Buying blanket-type insulation material is NOT recommended for gas water heaters. For safety, use ONLY the refit kit if you have a gas water heater.

Don't overheat your water. Use the lowest temperature you can. Most people find that 110° to 120° is good. Turn down the thermostat on your water heater. There is a dial for this, usually near the pilot light. Turn it from a "High" or "Hot" setting to a "Low" or "Warm" setting.

Use of Hot Water Heaters

Hot water heating, stoves, and refrigerators use up $1 out of every $5 that you spend for energy. Let's see if you can save some dollars here as well.

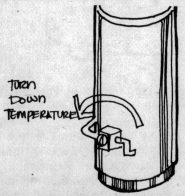

Insulate the pipe between the hot water heater and the faucet. This helps keep the hot water in the pipe warm. Either pipe insulation or wraparound insulation is good for this. You can buy either at most hardware stores.

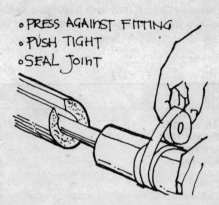

o PRESS AGAINST FITTING
o PUSH TIGHT
o SEAL JOINT

What else can I do to make household hot water use more energy-efficient?

— Lower the temperature setting on your heater to the lowest degree acceptable to your needs.
— Never leave hot water running unnecessarily.
— Fix leaky hot water faucets.
— Limit dishwasher runs to full loads.
— Consider installing a flow restrictor in the showerhead pipe to reduce water flow. Flow restrictors are available at most stores that handle plumbing supplies.
— Consider using faucet aerators. They mix air with water and reduce the amount of water used yet provide a water flow turbulent enough for washing.
— Launder clothes in cold or warm water whenever practical.
— Run full wash loads or adjust the water level control on your washer to the size of your washload.

ENERGY SAVINGS THROUGH AUTOMATIC THERMOSTAT CONTROLS

There is a myth that says you won't save energy by turning down your thermostat at night because it takes so much energy to warm the building in the morning. But this is untrue. Setting the thermostat back for several hours at a

Heating Costs Saved with Nighttime Setback
Approximate Percentage Saved with 8-Hour Nighttime Setback of —

City	5° F	10° F
Atlanta, GA	11	15
Boston, MA	7	11
Buffalo, NY	6	10
Chicago, IL	7	11
Cincinnati, OH	8	12
Cleveland, OH	8	12
Columbus, OH	7	11
Dallas, TX	11	15
Denver, CO	7	11
Des Moines, IA	7	11
Detroit, MI	7	11
Kansas City, MO	8	12
Los Angeles, CA	12	16
Louisville, KY	9	13
Madison, WI	5	9
Miami, FL	12	18
Milwaukee, WI	6	10
Minneapolis, MN	5	9
New York, NY	8	12
Omaha, NE	7	11
Philadelphia, PA	8	12
Pittsburgh, PA	7	11
Portland, OR	9	13
Salt Lake City, UT	7	11
San Francisco, CA	10	14
Seattle, WA	8	12
St. Louis, MO	8	12
Syracuse, NY	7	11
Washington, DC	9	13

stretch each day during the heating season—up, during the cooling season—will, in a centrally heated and cooled building, save energy. Depending on your geographical location, the amount of energy you can save will range from 9 to 15 percent of what you used before adopting this energy-conserving habit.

There are two ways you can accomplish temperature setback and setup: by adjusting the thermostat manually at the proper times or by installing a device that makes the adjustments automatically. The manual technique, of course, requires no special equipment, but it does demand a greater degree of time and attention than many people are willing to put forth day in and day out. An automatic control device, on the other hand, involves some initial investment, but this outlay is more than repaid in dependability and energy savings over a period of time.

How Much Will A Setback Device Save?

The exact energy and cost savings from a setback device are dependent on building design, amount of insulation, climate, temperature setting, and utility rate structures. Several studies have been conducted to estimate the fuel and cost savings that can be realized by using a setback device during both the heating and cooling seasons. The table on the left represents the *approximate* percentage of your heating costs that can be saved in various cities throughout the country for an 8-hour nighttime thermostat setback of 5°F and 10°F.

How To Estimate Costs Savings

From the table, you can estimate what you are likely to save by automatically setting back your thermostat during the heating season from 65°F to either 60° or 55°F at night. For example, if you live in or around Detroit and your heating bills amount to approximately $300 for the heating season, by lowering your thermostat temperature at night from 65°F to 55°F, you could save as much as 11% of $300, or $33 a heating season. These estimated figures are based on an assumed daytime setting of 65°F.

Types Of Automatic Controls

Two types of automatic controls are now available on

the commercial market. One is a device that works with a conventional thermostat. The other type requires replacing the existing thermostat.

Converter Setback Device: This type of control converts any existing thermostat to a timed device. Several variations are available. One is a two-component system in which a temperature control is mounted below the existing thermostat and is connected by wires to a separate timer unit plugged into a wall outlet. If the wires carry low voltage current they can be concealed in the wall, if desired. 110 volt power cords cannot be so concealed. Another is a single-unit device that is attached to the wall below the thermostat and is either plugged into a nearby wall outlet or operated by self-contained batteries.

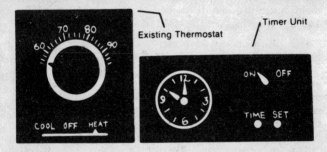

Existing Thermostat

Timer Unit

Replacement Setback Device: This type of control replaces the conventional thermostat entirely and is generally wired to the building's electrical system and heating/cooling system. Several types are available, but this type of device is usually more expensive to buy and costly to install, since it usually requires the additional wiring in existing walls. Its main advantage is that, having all wires hidden, it gives a neater appearance.

Automatic setback devices are sold in many hardware and department stores as well as building material outlets. In general, converter types sell for less than replacement types and can be installed by a do-it-yourselfer. Most converters retail for less than $40 whereas the initial cost plus the cost of installing a replacement unit may range from $75 to over $100 depending on the model and the type and extent of installation labor required.

INSTALL A HEAT METER

The heat meters have a wide range of application. For example, they are useful in measuring the heat consumption of individual apartments supplied by a central heating system so that charges may be levied based upon individual consumption.

The heat meter consists of a water flow sensor, two temperature sensors, electronic computer and register. It senses the flow of water and the hot and cold water temperatures, computing and accumulating the number of Btu's over an extended period of time.

Heat Meter — Single Family

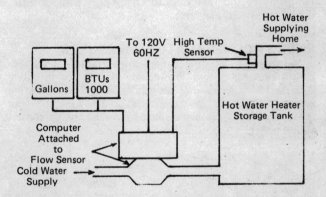

Applied to a domestic water heating system supplying one user as in the case of a single family, the meter with the computer mounted on top of the flow sensor is inserted in the cold water pipe supplying the hot water heater. The meter's low temperature sensor is already wired in and mechanically secured to an inside surface of the flow sensor.

The heat meter also measures the quantity of heat energy directly in Btu's which a solar energy or other type of heating system provides over a period of time. It is useful for determining heating system efficiency and making cost evaluations. The meter may be used to determine how the heat from a single system is distributed among many users and so provide a basis for equitably sharing system costs or for billing purposes.

The heat meter is also suitable for measuring the quantity of heat energy removed as with cooling systems and is available optionally for combined heating/cooling measurements.

A particular advantage of the meter is its high accuracy over a wide range of operating conditions. It automatically corrects for Btu register errors resulting from temperature induced changes in the characteristics of the flow sensor or the water itself. This feature permits heating system temperatures and flow rates to vary greatly without introducing significant Btu error.

ENERGY CONSERVATION MEASURES

A variety of tempting incentives are available to middle-class homeowners and low-income renters for installing such fuel-saving measures as insulation and solar heating equipment. Electric and gas utilities do play a major role in informing ratepayers about their individual energy conservation needs.

Utility Conservation Program

Under the National Energy Act, large gas and electric utilities are required to inform their residential customers about ways to save energy, suggesting the installation of appropriate equipment, and estimating possible energy savings from such improvements. They also must provide lists of businesses in the area that will finance, supply, and install the energy-conserving measures suggested under the bill. Also, utilities would have to offer to inspect each customer's home to advise him on his individual need for equipment or energy-saving practices to plug energy leaks. Utilities could act as a "project manager," or contractor, by offering to arrange for the installation and financing of residential conservation measures by other businesses and lending institutions. The customer could repay a lending institution for an energy conservation loan through payments on his regular utility bill.

Small Utility Loans

Utilities could make small loans of no more than $300 for the purchase or installation of specified conservation measures. The Department of Energy can expand this list or make modifications to reflect the varying climatic needs of different regions.

The conservation measures specified in the conference report include caulking and weatherstripping of doors and windows; new, efficient furnaces or boilers to replace old inefficient ones; ceiling, attic, wall and floor insulation; water heater insulation; storm windows and doors; special heat-absorbing or heat-reflective glazed windows; and solar wind-power equipment, including water heating, space heating, and cooling.

Three Devices Eligible for Higher Loans

If a residential customer is purchasing and installing one or more of the three special devices, a utility may make a loan for more than $300, if the cost of these measures is greater than $300. The three special types of equipment are (1) clock thermostats; (2) devices to increase the efficiency of furnaces, such as flue constrictors that limit the amount of heat escaping up the chimney, and electrical or mechanical furnace ignition systems that replace standing gas pilot lights; and (3) load management devices, primarily meters that measure how much energy has been used at different times of the day so that the utility can charge a reduced rate for off-peak power use.

Furnace retrofit devices alone could save the consumer 26% of his gas bill and 30% of his oil heating bill; these devices could save a significant amount of energy at a small expense. The utility can charge a reduced rate for off-peak power use. Utilities are prohibited from installing any conservation measures, except for the above three special devices. The utility program is directed at owners of single-family homes or very small apartment buildings (four units or less).

FIREPLACES AND HEATING STOVES

Both fireplaces and stoves can be used for heating the whole house, or you can use them for extra heat. When you have a choice between a fireplace and a stove, use the stove. Stoves give off more heat. If there are two chimneys in the house, use the one on the inside wall. It will draw better and give more heat, too.

Close the damper when the stove or fireplace is not in use. Be sure the fire is *completely* out! Leaving the damper open will waste a lot of heat. There are a couple of things you can do if your heater doesn't have a damper. On stoves, you can buy at many hardware stores a special section of stovepipe with a damper in it. If you have a fireplace, stuff a newspaper a little way up the flue. Pull the paper down and use it to start the next fire.

FIREPLACES

Many folks use fireplaces for extra heat. Usually a lot of wood has to be burned because fireplaces are not very efficient. This happens for two reasons. First, even small fires draw large amounts of cold air into the house to feed the fire. This creates cold drafts along the floor and cools the house. Second, as soon as the heat comes off the fire, it is drawn up the flue. Not much heat gets out into the room.

So a fireplace can even cool your house can even cool your house down! For the greatest comfort, close some doors and try to draw air in through empty rooms.

Don't build a roaring fire. These waste fuel because the heat goes up the chimney before you can use it. Burn wood and coals *slowly*.

The Fireplace Plug

It's best to use heavy-gauge, galvanized steel. Your *operculum* (or fireplace lid) should be large enough to extend over the opening so that it covers a portion of the surrounding brick. On the steel lid, the edges of sides and top are bent inward for a close fit against the brick. A handle is set on the front and steel shelf brackets serve as feet. The entire *operculum* front can be painted with black, heat-resistant paint.

Fireplace Air Vent

Put a vent under your floor to feed air to the fireplace. This will reduce the amount of air which is drawn across the room from cracks around windows and doors.

FIREPLACE AIR VENT

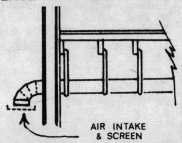

AIR INTAKE & SCREEN

The easiest way to connect a vent to the outside is to use a series of rectangular metal ducts. You can get these ducts at a plumbing supply store. Different sizes are available, but try to buy ducts which are at least 3'' deep by 10'' wide. You will need enough to stretch under your floor from just in front of the fireplace to the *nearest* opening in the basement.

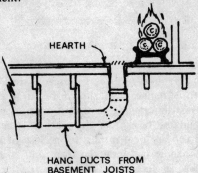

HEARTH

HANG DUCTS FROM BASEMENT JOISTS

Now, cut a hole for one end of the duct system right in front of the fireplace. Then make an opening in the basement wall or through a basement window for another duct to go through to the outside. When you put the basement duct in place, tilt it down a little toward the out-of-doors. This will allow any moisture to drain out.

Two elbows are attached to the duct system. One should be placed face down with a screen covering it, just outside the house where the basement duct comes through. This will keep the rain, wind, leaves, and insects out of the duct. The other is used where the fireplace duct comes through the floor just in front of the fireplace.

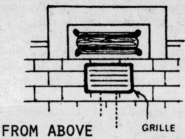

FROM ABOVE   GRILLE

Once you have the ducts in place, you finish the job by placing a grille over the opening in the floor. Use a strong grille so you can walk on it, and try to get one which can be closed. When the fireplace is not in use, close the grille and put a throw rug over it.

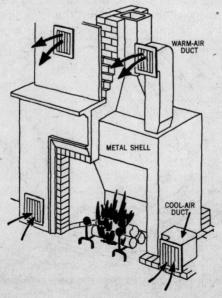

WARM-AIR DUCT

METAL SHELL

COOL-AIR DUCT

Fireplace Air Heater

The next step is to capture more of the fire's heat before it goes up the flue. There is an easy way to do this. Build a fireplace air heater which not only holds the burning wood but also forces warm air out into the room.

The heater is made of 1-inch iron pipes bent into a sort of "C" shape. The vertical pieces are bent around so that the top of the pipes extends just outside the fireplace.

Pipe elbows can be used to make the connection between these vertical pieces and the bottom pipe sections of the heater. Make the bottom pieces just long enough to come out in front of the mantel a few inches, but stay behind your new air vent grille.

Make enough of these "C"-shaped pipes to come all the way across the face of your fireplace, with a 3-inch or 4-inch space between each "C." The "C's" rest on and are attached to a base made of iron pipe also. The legs of

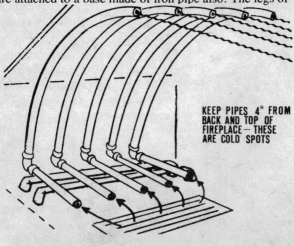

KEEP PIPES 4" FROM BACK AND TOP OF FIREPLACE— THESE ARE COLD SPOTS

AIR HEATING GRATE

the base should be 2'' to 3'' high and can be made by bending each end of the pipe. Have the "C"-shaped pipes welded to the base at a local welding shop. The cost should only be a few dollars.

This air heater works by sucking cold air from the floor. The air is heated by the fire, and then flows out into the room. It is possible to couple a fan along the front of the "C"-shaped pipes in order to increase the cold air supply and thereby increase the warm air circulation.

Install a wood stove

Adding a small stove can be a low-cost way of getting some extra heat. Wood and coal stoves, some in really great condition, can be bought cheaply at garage sales and auctions. Generally, it is worthwhile to purchase the good air-tight stoves as they hold a fire, are efficient, and do not create a fire hazard. Cover the rug or floor under your new stove with an asbestos sheet, and place tin or aluminum on top of the asbestos. This reflects heat up and keeps the floor cooler. Floor covering should be at least ¼'' asbestos millboard covered with sheet metal. Cover floor 12'' around sides and rear and 18'' in front.

Setting Up a Stove in a Fireplace

1. Measure fireplace and cut a sheet of metal or asbestos board to close the opening.
2. Set stove on a fireproof stove board at least one stove pipe length (24'') away from the face of the fireplace.

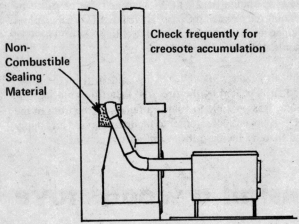

Non-Combustible Sealing Material

Check frequently for creosote accumulation

3. Install a damper in one section of the pipe.

4. Connect pipes to stove and elbow and adjust to determine proper location of hole in sheet metal or asbestos board.

5. Slide collar on pipe and insert pipe into the hole in closing board. Then fit the collar snugly against the board.

Caution: Burn coal only in a cast iron stove with a grate designed specifically for coal.

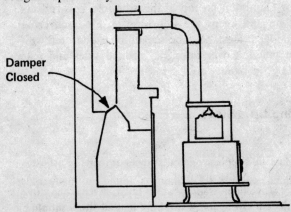

Damper Closed

SPECIFIC PROCEDURE FOR
INSTALLATION OF WOOD BURNING STOVES

Stoves must be set up and used with great care to avoid serious fire hazards. Safe chimneys are absolutely essential. Flue walls must be sound as occasional chimney fires are almost inevitable when burning wood or soft coal. Safe placement of stoves and proper vent connections are also important.

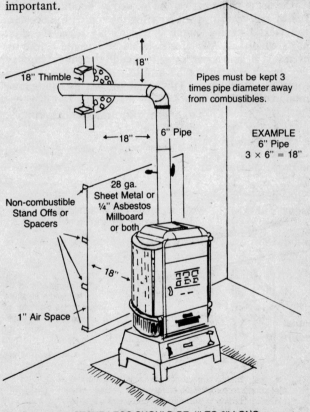

18"

18" Thimble →

Pipes must be kept 3 times pipe diameter away from combustibles.

← 18" →   6" Pipe

EXAMPLE
6" Pipe
3 × 6" = 18"

28 ga.
Sheet Metal or
¼" Asbestos
Millboard
or both

Non-combustible
Stand Offs or
Spacers

18"

1" Air Space

**ALL STOVE LEGS SHOULD BE 4" TO 6" LONG.
STOVE MUST BE ON A STOVE BOARD***

Unlined single brick chimneys found in many older homes are especially hazardous. This type of chimney often was not very safe when it was built and certainly should be suspect now. Mortar in the joints probably has broken down and some bricks may be cracked. The combined action of weather and hot gases causes these conditions most often near the chimney top. However, cracks and openings commonly develop well below the roof in tinder-dry attics.

Cracks in chimneys can be located by building a smudge fire in the bottom, then covering the top with a board or wet sack. Escaping smoke should readily reveal the chimney's condition. All defects should be repaired before use even if it becomes necessary to rebuild the whole chimney. Woodwork should not be in direct contact with the masonry of any chimney. This condition was also quite common in old construction.

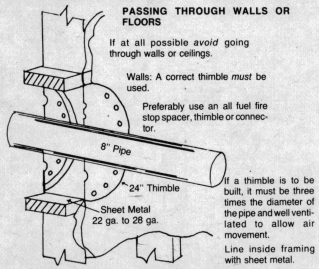

PASSING THROUGH WALLS OR FLOORS

If at all possible *avoid* going through walls or ceilings.

Walls: A correct thimble *must* be used.

Preferably use an all fuel fire stop spacer, thimble or connector.

8" Pipe

24" Thimble

Sheet Metal 22 ga. to 28 ga.

If a thimble is to be built, it must be three times the diameter of the pipe and well ventilated to allow air movement.

Line inside framing with sheet metal.

Ceilings: When you must pass through a ceiling, the all fuel connector or fire stop spacer must be used and installed according to the manufacturer's requirements.

Fireplaces and stoves, when fired vigorously from day to day, were usually not as hazardous as the controlled-burning stoves common today. Soot and creosote did not build up as small accumulations may have been ignited and burned safely more or less continuously. Heavy chimney deposits, once ignited, burn intensely at dangerously high temperatures.

CONNECTING THE STOVE PIPE TO A CHIMNEY, THROUGH A WALL

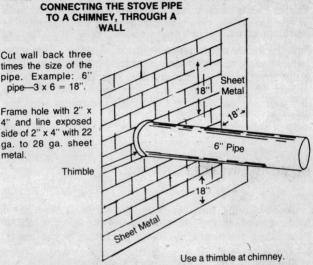

Cut wall back three times the size of the pipe. Example: 6" pipe—3 x 6 = 18".

Frame hole with 2" x 4" and line exposed side of 2" x 4" with 22 ga. to 28 ga. sheet metal.

Thimble

Sheet Metal

18"

18"

6" Pipe

18"

Sheet Metal

Use a thimble at chimney.

Present-day building codes and insurance underwriters encourage safe chimney design. Masonry flues are lined with fire-clay at least 5/8-inch thick or some other approved material. All wood beams, joists and studs must be kept at least 2 inches away from masonry enclosing a flue. Approved, factory-built chimneys, when correctly installed, are also acceptable.

A minimum clearance of 18 inches is recommended between combustible materials and single-walled metal pipe connectors (stove pipe). Combustible surfaces may be protected in an approved manner to allow closer spacing when necessary. For example, 28-gauge sheet metal that is spaced out 1 inch from a wall or ceiling on non-combustible spacers permits a 9-inch clearance. Also, surfaces can be protected by a 1-inch layer of wire mesh, reinforced, non-combustible insulation covered with 22-gauge sheet metal to allow a clearance of only 3 inches. It is also recommended that the horizontal length of uninsulated stove pipe to a chimney be not over 75 percent of the vertical portion of stove pipe. Unnecessary turns should be avoided.

Space heater placement varies depending on whether the heater is a jacketed, circulating type or radiating type. Twelve inches of clearance is needed at the sides and back of the circulating type and 36 inches is recommended at the sides and back of radiating types. Both types should have 36 inches of clearance overhead, 4 inches of space between the stove bottom and a protective floor cover of 24-gauge sheet metal or its equivalent. This floor cover should project 12 inches beyond the stove on the sides and back and well beyond the front for the safe removal of ashes and servicing.

BUYING A WOOD FURNACE

Wood-burning furnaces come in two basic types. One is the forced-air type that distributes hot air throughout your home by means of ducts. The other is the boiler, for hot water heat.

Building a new home? Either system works well, but forced air has two advantages: add-on humidification and air conditioning. If you are buying a woodburner to replace an existing furnace, buy the kind that matches the present system.

Multi-fuel or wood only? Multi-fuel furnaces can burn both wood and gas or oil. They save you money when you feed them wood. But when you aren't home to stoke, they keep the house warm by burning fossil fuel.

Another way is to achieve the same effect by installing a wood-only furnace in tandem with an existing gas or oil furnace. Set the thermostat on the gas or oil furnace a few degrees lower than the thermostat on the wood burner. Then whenever the wood furnace dies down, the other furnace takes over. For safety, you shouldn't connect the two furnaces to the same chimney flue. An extra chimney for the wood furnace can cost a lot of money. So it might make more sense to sell your existing furnace and buy a multi-fuel unit, which needs just a single flue. Overall, this will probably be the cheapest way to go, and it will save space in the basement over a two-furnace installation.

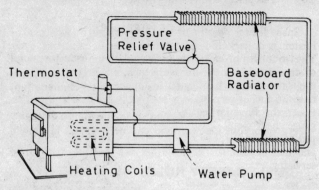

What kind of maintenance is needed? Some require a fairly careful cleaning every few weeks. Is the wood fire started automatically, or do you have to kindle it up yourself? How big a log will the furnace take? Longer logs mean less work than shorter ones. And how long will the furnace run on a single loading? Most will go about 12 hours. If you are considering a multi-fuel furnace, what fuel besides wood does the furnace burn? Some manufacturers offer only wood/gas furnaces, some offer wood/oil, and others offer a multi-fuel combination. The best way to compare one furnace against another is to read the owner's manuals and the sales literature.

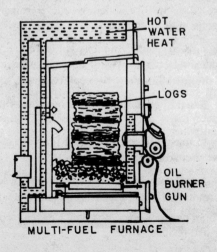

MULTI-FUEL FURNACE

Free Standing Stove or Fireplace

1. Set stove on fireproof foundation (asbestos-metal stove board, brick, marble, chips, etc.).
2. Install a prefabricated, insulated metal chimney, approved by the Underwriter's Laboratories (UL).
3. Prefabricated chimneys are available for use through a roof or an outside wall and can be installed in mobile homes.

SIMPLE SAFETY RULES

DO—Make sure of proper clearances from combustible floors, walls and ceilings.

DO—Have the chimney inspected by a competent mason.

DO—Check the condition frequently for signs of deterioration.

DO—Use a metal container to dispose of ashes outside the home.

DO—Install a smoke detector and fire extinguisher.

DO—Use, dry, well-seasoned wood. Burning green wood results in dirty chimneys.

DO—Be sure to check your city or county fire and safety codes. Also, inform your home insurance company of the installation of a fireplace.

DON'T—Extend pipe through walls or ceilings if at all possible.

DON'T—Connect a wood stove to a fireplace chimney unless the fireplace has been sealed off.

DON'T—Use flammable fuels to ignite wood. Use paper or kindling.

DON'T burn pine or other wood which develops pitch or other tar-like residue.

DON'T—add an extra flue to your chimney without checking with your local fire department or building official.

Fuel your stove

You can burn just about anything in "chunk stoves." You can burn paper, logs, small chunks of wood, broken-up wooden boxes, or coal. A nice way to make use of extra newspaper or magazines is to make paper logs. Roll up the paper in round log-like shapes, tie these with string, and let them soak in water until fully wet. Dry them out near a heat source. The moisture helps fight winter dryness in the house. Once dry, paper logs burn almost like wood and keep a good fire going for quite some time.

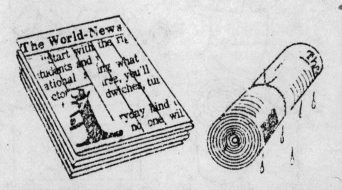

Use Flue Heat

On both furnaces and heating stoves much of the heat goes up the flue and is wasted. You can take some of this back by putting metal "donut rings" on the flue. These give off heat.

The easiest and least expensive metal to work with is aluminum. You can buy "aluminum flashing" or "aluminum grass edging" at a hardware store. Get one roll of it.

The donuts in place on the flue

To get more heat, make sure the flaps are staggered

FLAP

FLAP

ALUMINUM GRASS EDGING

COLLAR

Get all the heat you can from the fuel you burn! Keep your heating system in top shape and burn the fuel carefully and slowly. You can save another $10 out of each $100 you usually spend for heat.

Heating with Wood

Increasing costs and, in some cases, current or projected shortages of other fuels have increased interest in the use of wood as a heating fuel. The manufacturers of wood burning units have responded to the new interest in fuelwood with improved construction and efficiency in their combustion units.

Fuelwood is either sold by weight, the load, or the stack. There are various sized stacks and confusion often exists over the amount of wood in a "cord." A standard cord of wood is defined by law as a pile 4' high and 8' long made up of sticks 4' in length. A "face cord" is a pile 4' x 8' made up of sticks of any length (often either 12", 16", or 20"). The amount of solid wood in the stack depends upon the size and straightness of the sticks and whether they are round or split. Thus, a standard cord may contain from 60 to 110 cubic feet of solid wood.

Wood dealers often eliminate the need to define a cord by selling wood by the load or by weight. Of course, the amount of wood in a "truck load" depends upon the type of vehicle.

Fuelwood Preparation

Some of the other factors that should be considered when preparing wood for burning are:

   Cutting — Should be done at least six to nine months prior to burning. It often requires a chain

Standard Cord

Total Volume = 128 cu. ft.

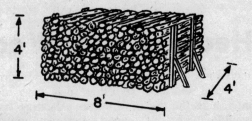

Face or Short Cord

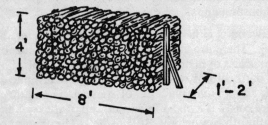

saw which is perhaps the most hazardous operation connected with preparing wood. Keeping the leaves on summer-cut trees until they wither helps remove a great deal of moisture from the wood.

Splitting — Greatly reduces drying time. It is often necessary for efficient handling and combustion and is best done when wood is frozen or green.

Stacking — Necessary for proper drying of the wood and should be done immediately after splitting. Cover and allow for adequate air circulation.

Seasoning — Necessary to reduce moisture content of the wood and assure proper combustion.

How Wood Burns

Wood burns in three phases — (1) Heat drives water from the wood. (This heat does not warm the stove or the room.) (2) Charcoal and volatile gases are formed. The gases can produce 50 to 60 percent of the heat value of the wood; but they must be heated to about 1100°F and mixed with sufficient oxygen to burn. (3) Following the release of the volatile gases, the charcoal burns. These phases overlap so that all occur at the same time.

Fuelwood Characteristics

The heat derived from the combustion of wood depends upon the concentration of woody materials, resins, ash and water. In general, the heaviest woods (hickories, oaks, locust), when seasoned, have the greatest heating value per cord. Lighter woods (aspen, basswood, willow) give about the same heat value per pound but they give less heat per cord because they are less dense. When considering the type of wood to burn, other important characteristics are:

—ease of splitting (apple, birch, maple, oak)
—ease of ignition (birch, cedar, pine)
—production of heavy smoke (cedar, spruce)
—sparking (cedar, hemlock)
—coaling qualities (apple, cherry, hickory, maple, oak)

The use of wood for home heating has several disadvantages. Fuelwood must be well-seasoned (dried) in order to be most efficient.

As a fuel, wood burns rapidly so that refueling must be frequent. Also, it is heavy and hard to transport.

Although the production of fuelwood is a rather dirty business, wood is a relatively clean fuel

Variation of Heating Values of Wood Due to Moisture

Percent of Moisture	Percent of Usable Heat
0 (oven dry)	103.4
4	102.7
10	101.6
20 (air-dried hard wood)	100 (7,250 Btu)*
40	96.5
80	89.7
100	85.00

*Btu is the quantity of heat required to raise the temperature of one pound of water one degree Fahrenheit.

producing about 1 percent ash by weight. Wood heat produces relatively small amounts of chemical air pollution. But the production of particles in the air tends to be higher than with other conventional fuels.

A disadvantage in heating with wood is the high fire hazard. To prevent the possibility of fire outside the heating unit, the wood should be burned in sound, well-constructed stoves or fireplaces. Stoves should have dampers to control the rate of burning. The chimney must be sound with all joints properly mortared. When constructing new chimneys, use earthen flue tile to prevent super-heated gasses from escaping between the bricks to flammable materials.

Heat Values

Generally, heat values are dependent on the percent of moisture and the weight of the wood. Heavier woods have a higher heat value. The heat value from an air-dried standard cord of several native hardwoods (such as hickory and oak) when burned in a modern efficient woodburning unit is equal to nearly 130 gallons of No. 2 fuel oil.

Burning characteristics vary with wood species. Elm tends to burn slowly with little or no flame while white birch and pine burn quickly with much crackling and spark-throwing. Green wood is not efficiently burned in ordinary stoves.

The lighting or kindling of a fire in a stove or fireplace is dependent upon heating the wood to the point of ignition or the kindling point. Some woods also have a low kindling point, such as small pieces of dry white pine; pieces of white birch bark or softwood cones can be used to ignite woods that are more difficult to burn. Also, paper and cardboard are good kindling materials. Highly flammable liquids such as gasoline should NEVER be used to kindle fires.

Approximate Weight and Heating Value per Cord (80 cu. ft.) of Different Air-Dried Woods

Woods	Weight lb. Air dry	Avail. heat Millions Btu	Equivalent in gallons of fuel oil
Ash	3440	20.01	145
Aspen	2160	12.5	91
Beech, American	3760	21.8	158
Birch, Yellow	3680	21.3	154
Elm, American	2900	17.2	125
Hickory, shagbark	4240	24.6	178
Maple, red	3200	18.6	135
Maple, sugar	3680	21.3	154
Oak, red	3680	21.3	154
Oak, white	3920	22.7	165
Pine, eastern white	2080	13.3	96

CHIMNEY CONSTRUCTION

Chimneys are constructed of either masonry or prefabricated metal. The metal chimneys have concentric walls with air spaces or insulation in between. The chimneys should have the label ALL FUEL from a recognized testing lab such as Underwriter's Laboratories (UL). Masonry chimneys may be brick, cinder block, or stone. Tile flue liners are standard for masonry chimneys. Older chimneys often have no tile lining so check them carefully for leaks. It is best to locate the chimney on an interior wall to maintain higher flue temperatures and thus to reduce the formation of creosote. The figure illustrates one approved method for connecting the stove pipe to the chimney. The cost of metal pre-fab versus masonry chimneys depends a great deal upon the individual installation method used. Masonry chimneys usually have the longest life. Pre-fab chimneys with a stainless steel inner and outer lining have a longer life than those of galvanized sheet steel.

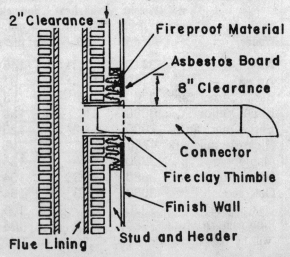

Multiflues

Do not connect a wood stove to the same flue serving a fireplace because sparks and flue gases from the stove may enter the house through the open fireplace.

Room heaters, cook stoves, etc. should not be connected to a common flue because (1) flue gases and sparks may pass from one flue opening to another and (2) multiple connections sometimes cause a poor draft and unsatisfactory operation. If, despite these recommendations, two stoves are connected to the same chimney, the connections must enter the chimney at different elevations.

WOOD STOVE SAFETY— THE CREOSOTE PROBLEM

Creosote is found almost anywhere in a wood heating system, from the top of the chimney to the inside of the stove itself. It is caused by unburned gasses found in wood smoke which condense on cool surfaces.

The form which creosote takes depends on the temperature of the surface on which it condenses. For example, if it condenses on a relatively cool surface, such as an exterior stovepipe, the creosote will contain much water and will be very fluid. It may even be seen dripping from the joints of the stovepipe.

Creosote-clogged Pipe and Flue

When there is condensation on a surface of 150°F or more, a smaller amount of water is present and the creosote will be very thick, sticky and similar to tar. This form is particularly hard to remove from surfaces. Whatever form creosote takes, it is always dark brown or black and has a very unpleasant, acrid odor.

Hard to Remove

If allowed to remain in the chimney or pipe, the form of creosote will continue to change. The longer it is heated, the more water will evaporate, until finally the creosote takes the form of carbon. In this form it is flaky and shiny on one side and may be brushed or scraped off. Other forms are difficult to remove even with a stiff wire brush.

Many factors influence creosote buildup. Probably the most commonly-discussed factor is related to the type of wood burned and its moisture content. Dry hardwoods are generally assumed to generate the least amount of creosote but the quantity can still be large. Creosote formation is not entirely eliminated no matter what kind of wood is burned.

The amount of creosote deposited depends mostly on (1) the density of the smoke and vapor from the fire (the less smoke, the less creosote), and (2) the temperature of the surface on which it is condensing (the higher the temperature, the less creosote.

Smoke Density Factor

Creosote generation is highest during low, smoldering burns. Smoke densities are least when combustion is relatively complete. This tends to be the case when the amount of air admitted to a wood burner is high. For this reason, leaky stoves, open stoves and fireplaces usually have fewer creosote problems than other wood-burning devices.

Smoke density can be lowered somewhat in an air-tight stove by using small amounts of wood and stoking more often or by using larger pieces of wood. Creosote formation can be limited by leaving the air inlet slightly open after adding wood to promote more rapid burning until the wood is mostly reduced to charcoal. Then close the inlet as much as desired. This causes more complete combustion and burns the potential creosote-forming gasses. An additional amount of heat will be generated while the gasses are burning. This is a house-warming bonus.

Can Cause Chimney Fires

Because creosote forms more quickly on cooler surfaces, a well-insulated pre-fabricated metal chimney has the least serious creosote problem. The insulation helps keep the temperatures of the inner surfaces higher and the chimney's low heat capacity lets it warm up more rapidly after a fire is started. Flue temperatures can be increased by using a shorter stove pipe to connect the stove to the chimney. However, this decreases the energy efficiency of the system.

Creosote can burn and cause potentially-dangerous chimney fires. If a fire starts, the amount of air entering the stove should be decreased as much as possible.

Creosote, because of its acidity, causes corrosion in many materials, including steel and mortar. Masonry chimneys with tile liners or pre-fabricated insulated chimneys with stainless steel liners are corrosion-resistant. When properly installed, both types are also safer than non-lined chimneys in the event of a chimney fire. Chimneys without liners which are poorly-maintained are hazardous.

Keep Chimney Clean

In summary, to lessen the formation of creosote:

1. Keep the chimney clean. Check the need for cleaning before starting a fire in the fall. Then check after two weeks, one month, two months, etc., until you can determine how frequently your chimney should be cleaned.
2. When you are sure that your chimney is **safe** and **clean,** run a hot fire at least once each day, preferably in the morning while there is someone available to watch it.
3. Burn dry hardwood when possible.
4. Leave the air inlet slightly open after adding wood to promote more rapid burning until the wood is mostly reduced to charcoal. Then close the inlet as much as desired.
5. Use small amounts of wood and stoke more often or use larger pieces of wood.

Section 7

HOME HEATING
IN AN EMERGENCY

At some time a situation may develop in which you might face a heating emergency—when your home heating system is inoperative for hours or days. At the critical time you must decide how to meet the emergency, either with an alternative source of heat or by seeking shelter elsewhere. Serious planning and preparation should be considered.

Safety is of prime importance in choosing an alternate form of heat. Consider all potential hazards and eliminate as many as possible, keeping in mind that your degree of protection is lower during a community emergency. Normal community services such as police and fire protection, doctors, hospitals, and highway maintenance may be in great demand and unable to respond to your emergency immediately. Under emergency conditions you may have to do certain things you wouldn't consider doing under normal circumstances. Use extreme caution.

Preparing for an Emergency

The first step in making a plan is to determine the conditions your family might face if your heating system fails. Because all members of your family would be affected, each should have a hand in the planning. Try the following:

Discuss with your family what you might do if the heating system went off and were to remain off for several days and nights.

If your home is heated electrically, failure would obviously be caused by lack of power. But don't forget that most other systems depend on electricity, too. Oil burners usually have electrically-powered fuel injectors and ignition. Hot-air systems rely on a fan for air circulation; hot-water systems with zone valves and circulator pumps, or coal furnaces with motorized stokers, also need electricity. Most thermostats require electric power to operate.

Imagine that your area is experiencing an intense storm. It is cold and telephone service is disrupted. Then, with a pencil and pad handy, discuss how you would cope with the crisis. The family would have to determine what could be done to provide home heat, or at least how to keep members of the family warm. Discuss sources of alternate fuels available; how to get them and how to use them; what protective measures would be necessary such as keeping pipes from freezing; and supplying water if the pump is not operating. As part of the discussion you probably will want to draw up a list of additional obstacles that might be encountered, the responsibilities of each family member, and supplies available.

Where Do I Begin My Planning?

First, consider the resources you now have in your home for meeting emergencies. Because no two homes are the same, each homeowner must assess his own situation and prepare accordingly. Keep that pencil and paper handy!

Your Resources

A. Could your heating system, with simple modification or through manual operation, continue to heat all or part of your home?

B. What other heating devices are used or stored in your home, garage, or barn? List them. Some suggestions:
   Fireplace
   Charcoal grill
   Wood, coal, gas, or oil stove or space heater
   Camping stove or heater
   Electric or gas oven and surface heating units
   Portable gas oven
   Gas-fired hot water heater
   Portable electric heater

C. List fuels available in your home or within reasonable distance. Which of them could be used in the above list of devices?
   Oil or kerosene
   Furnace, stove, or cannel coal
   Firewood, lumber, scraps, corncobs, straw
   Gas, campstove fuel, charcoal, starter fluid, alcohol, gasoline, motor oil
   Newspapers, magazines

If your heating device and fuel can be matched, would they provide enough heat to warm at least one room in your home? Is there enough fuel for several days? Do you have a secondary source of emergency heat?

Decide Now

If your regular heating system cannot be modified for an emergency, consider buying, building, or adapting a device or system that will. The choice might be a space heater, castiron or sheet-metal stove, or a catalytic heater. A small generator might be the answer if it will keep your furnace in operation. Your supplier or County Agent can help you decide what capacity generator you need. Perhaps you have been looking for an excuse to finally build that fireplace you've always wanted. Try to avoid depending on the same fuel for emergency heat as you have in your normal heating system.

Preparation

Now that you have decided how to heat your home during an emergency, it is time to get busy making preparations. Good planning now will give your family the confidence it needs when an emergency arises.

You will probably have to make some changes in your home or in your heating system to accommodate another heating device. If you can't make them, call in someone who can. Any device which burns fuel must be vented outside the house—both to eliminate smoke and gas and to provide oxygen for combustion.

Altering Your Regular Heating System

Minor alterations to your regular heating system might be considered:

A. Because automatic heating systems are often dependent upon electricity, you might wish to consider an emergency generator to provide power for full operation. This applies only to fossil-fueled systems with pumps, blowers, circulators, fuel injectors, electric ignition and thermostats. Electrically operated valves in many steam or hot air systems can often be operated manually. Hot air systems, depending on installation, are capable of providing limited heat without a blower. A coal-burning furnace can be fired the old fashioned way—with a shovel. Most small electrical generators such as might be available to the homeowner can supply only very limited power, inadequate for heating in an electrically heated home.

B. Sometimes another type of fuel can be burned in a heating system, for example wood can be used in a coal furnace. Get to know the capabilities and options of your primary heating system. If it can function at least partially in an emergency, it is your best source of heat.

Providing Vents and Flues

A. Install a "thimble," (a metal pipe which is inserted through the side of the chimney into the flue) to allow hooking up of a stove or space heater (See figure 1). If the heating device will be connected only during an emergency, fit the thimble with a metal or asbestos cap to cover the hole.

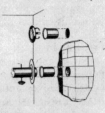

Note: chimney flues are designed to accommodate a single heating device at a time. Using more than one heating device at the same time on the same flue passage may result in smoke damage and improper burning of the fuel. If your auxiliary heating unit is to remain attached to the flue being used by the furnace, fireplace, or other burner, it should be fitted with a damper which will close off the device. Gas flues, which are usually smaller and lighter, cannot safely accommodate oil, coal, or wood burners. Gas devices, however, can be hooked to oil, coal, or wood flues.

B. Fireplaces in some homes are designed for appearance, not for their heat producing ability. If yours doesn't heat well, plug the throat with a piece of sheet metal with a hole cut for a stovepipe. (See figure 2.)

In an emergency, a stove or heater can be set on the hearth. Stoves are better, more efficient heat producers than fireplaces.

C. Conventional masonry fireplaces are often not efficient producers of heat and may take more heat from a room than they put in. Heat circulating or "heatilator" fireplaces are much more efficient. Their ease of installation may offset their initial higher cost when compared with construction of conventional masonry fireplaces. Also, a glass-doored, heat cir-

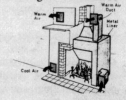

culating fireplace with special outside air inlets makes
a satisfactory heater that can use wood, coal, and other
combustibles.

D, If your present chimney cannot be used with an auxiliary heating system, consider installing a prefabricated
chimney for use with your alternate source of heat.

Using Other Fuels

A. If oil is your emergency heating
fuel and you have an oil fired
furnace, install a drain cock or
valve in the fuel line to draw
oil from the tank. A siphon hose
might be used if the tank has an
access plug.

B. An emergency generator to keep
your heating system functioning
will involve special wiring for
the changeover from utility power. Have an electrician
advise you on this.

C. If gas is the standby fuel, be sure to have proper fittings,
tubing, and tools on hand for a quick, safe hookup or
changeover. Be sure that before placing any valve on
your oil line that it is approved by your local building
authority. Consult your fuel supplier on this change.

D. Heat pumps, units similar to air conditioners, can supply
considerable amounts of heat under certain conditions.
They may be practical for your situation as a source of
heat which requires only electricity.

There is considerable heat in well water, which is
usually at least 50 degrees F., in the northern States. Depending on the depth of water, this may be an efficient
source of heat. The heat pump removes heat from the
water and transfers it to the home. In warmer areas
heat can be removed from outside air. Your local heating
or air conditioning contractor can help you decide if
a heat pump is practical for your situation.

Generators for Emergency Power

An electric generator could supply power to run furnace
blowers and oil burners and some other appliances in time
of emergency. Just how many appliances you could operate

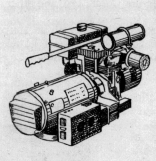

depends on the output of the generator. Before buying a generator, the homeowner should add up the wattage required. Motor requirements should be figured at their starting rate (much higher than the running rate) to arrive at the total number of watts required at peak use. Generators are rated according to their kilowatt output (a kilowatt equals 1,000 watts).

Additional costs would be necessary to rewire the home service entrance, to install a transfer switch, or to add an alarm device or other accessories as desired, and for regular maintenance of the standby system. Home generators are usually driven either by an attached gasoline or gas-powered engine or a portable power source such as a tractor. The best information on a generating system for your home can be obtained from a local supplier, your utility company, or your County Agent or civil preparedness representative.

Conserving Heat

1. What other materials exist that could be used for conserving body warmth or emergency heat? Winter clothing, especially bulky items and outdoor garments, sleeping bags and small tents, blankets and bedding, drapes, curtains, slipcovers, rugs, large towels, etc., should be considered.

   Remember, if all else fails (and you can't get to other shelter) bed is the warmest place to be, with other family members and lots of covering.

2. How much of your house should you attempt to heat? When the heat goes off and you are going to have to rough it, the smaller the space you heat the easier the job will be. What you do will be dictated by the amount of emergency heat you have available, the floor plan of your house, and the severity of the cold outside. If you will be utilizing your fireplace or a stove requiring a chimney flue, the choice of rooms has been made for

you. If, however, you will be able to obtain some heat from your furnace, select an area near it to cut down on heat loss that occurs in long pipe or duct runs. If you plan to use a portable heating device or have a choice among several heating zones, select an area on the "warm" side of the house away from prevailing cold winds. This area should have good insulation, as few windows as possible to minimize heat loss and should be capable of being isolated from other unheated areas either by closing doors or blocking openings to prevent drafts and heat loss. You may want to hang blankets or heavy drapes over windows to further reduce heat loss.

If you will be using your furnace in an emergency, know in advance how to prevent it from sending heat to unnecessary areas. In addition to shutting off the thermostat, this may involve blocking hot air ducts or shutting off certain steam or hot water lines.

Storing Emergency Fuel

Obtain fuel for your alternate heating system and store enough to last several days. Store it in a safe, convenient place such as a garage, carport, or shed away from the house. Do not use your emergency fuel for any other purpose, and check the supply regularly.

Community Resources

What resources are available for emergency assistance in your community? There may be town, school, or county plans for coping with emergencies. Your local Red Cross or civil preparedness authorities may have contingency plans and supplies. Find out.

A. Are there stockpiles of fuel available such as coal, oil, or firewood? (Some towns keep emergency supplies of firewood on hand at dumps or highway department sheds. If yours doesn't, perhaps it should.)

B. Are there any emergency supplies of foodstuffs and water? A civil preparedness representative or your County Agent could advise.

C. If your family were forced to leave its home, where could it go? Under what conditions? Schools and municipal buildings often have emergency lighting equipment and heat.

D. You may want to consider a cooperative emergency plan which combines your resources with those of a neighbor.

Related Heat Loss Problems

Keeping your family warm obviously won't be the only problem you will face if an energy failure strikes your home. Consider the following:

Freezing Pipes

If the heat will be off several hours or more and the temperature well below freezing you will have to protect exposed plumbing. Drain all endangered pipes, including hot water heating pipes in rooms that will not receive emergency heat. Familiarize yourself with your home plumbing and heating layout in advance so you can do the job quickly and thoroughly to avoid repairs later.

It may be necessary to install additional valves to enable you to drain only portions of your system. Don't forget the sink, tub, and shower traps; toilet tanks and bowls; your hot water heater; dish and clothes washers; water pumps, and your furnace boiler, if you have one.

Water for Household Use

If you rely on electricity to run your water pump, a power outage could restrict your water use. Save as much water as possible while draining your system and store it in closed or covered containers, preferably where it will not freeze. In addition to water in pipes, a sizeable amount can be collected from your hot water heater if you have one, and toilet storage tanks. Water from the heating system may be unfit for drinking or other household use.

Lighting

Have a good supply of candles, matches, and at least one kerosene or gas lantern with ample fuel. You should have a dependable flashlight with spare bulbs and batteries. If any of these materials are used when there is no emergency, they should be immediately replenished.

Sanitary Facilities

If your water supply is shut off, sanitation will become a problem. Disconnect the chain or lever attached to the toilet handle to prevent accidental flushes and instruct users to put toilet paper in covered containers. Flush only often enough to prevent clogging. An alternative might be to purchase a portable camper's toilet.

Emergency Cooking

During an emergency, providing hot meals for your family may be a problem. A camp stove can be used or, if necessary, cooking can be done in a fireplace. Keep a suply of meal-in-a-can foods such as stews, soups, canned meats, beans, or spaghetti to supplement dry stores like cereal, bread, dried meats, and cheeses. Freeze dried meals used by campers and backpackers are often excellent foods which can be prepared with a minimum of heat.

Safety

Review all your plans and preparations to ensure the safety of your family. Emergency actions are of little value if they lead to a new or bigger emergency. If you don't already have them, a good fire extinguisher and first aid kit are MUSTS!

For the Unprepared

Your home heat is gone. You've just discovered your system has stopped functioning and may be off for several days. It's cold outside, the temperature is dropping on the inside, and you have a first class emergency on your hands. What can you do?

Your first concern should be to conserve body heat. Keep the people in your household warm while you provide emergency heat. The simplest solution to this problem is to put on suitable clothing, or perhaps get into bed.

Safety is of paramount importance in a heating emergency. Few (if any) Americans have frozen to death in their homes in recent times. Many have perished from burns, smoke inhalation, or carbon monoxide poisoning. Loss of home heat constitutes an emergency, but it needn't result in tragedy.

Handling a heating emergency, once immediate requirements for body heat are met, can be broken down into five steps.

1. Finding a heat source or improvising one.
2. Obtaining fuel.
3. Selecting a room or area to be heated.
4. Setting up, testing, and operating an emergency system.
5. Dealing with related problems caused by heat loss.

In an Extreme Emergency

In an extreme emergency you may have to use a makeshift heating system as illustrated below. Unless a makeshift chimney has been inspected by your local Fire Department and approved by your Insurance Agent you will probably be violating your standard chimney warranty in your Fire Insurance Policy.

Selecting an Alternate Heat Source

What kinds of heating devices do you have which use readily-available fuel such as wood, coal, electricity, gas or oil? Perhaps you have a space heater you have used in your home, workshop, or shed; perhaps a stove or an electric, gas, or oil heater. Do you have a camp stove? Don't overlook the oven in your gas or electric range. If the fuel is available, turn the range on and open the oven door.

Stoves should be connected to a chimney flue if at all possible. Many older homes have capped stovepipe thimbles in rooms once heated by stoves. Another possibility would be to remove the nonfunctioning furnace pipe from its flue entrance and hook up your stove or heater in its place.

Sometimes a stove pipe can be extended through a window to provide proper venting of gases. **Note**: If no chimney exists, or cannot accommodate a thimble, or the building design prevents your using a prefabricated chimney, a window can be altered to provide proper venting. Replace the glass with a metal sheet through which a temporary stove pipe can be run outside the home. BE SURE

NO HEATED SURFACES ARE CLOSE TO THE SASH OR OTHER FLAMMABLE MATERIALS.

1. When setting up emergency stove piping, be careful about running it too close to flammable materials. This is particularly true when using a window-mounted flue. The wood sash is flammable as are curtains or shades that might normally be on the window.
2. Flues and piping for gas-burning appliances are designed primarily to vent vapors and may be unsafe for use with higher temperature oil, coal, or wood smoke.
3. A damper in your emergency flue will help facilitate satisfactory burning and regulation of the heat. Cutting down an excessive draft helps keep the heat in the room and prevents the flue from over-heating. Close the damper as far as possible without reducing combustion or forcing smoke into the room.
4. Natural gas appliances will not burn bottled gas, even in an emergency, without a mechanical conversion. Your local gas supplier has the materials needed for conversion.
5. If you use a catalytic or unvented heater provide plenty of ventilation. Keep a nearby window open at least one inch whenever the device is in use.

The least desirable solution—but any heat may be better than none—is to rely on a system utilizing a make-shift heater, including campstoves, stackless kerosene space heaters, or industrial-type oil or kerosene jet heaters. *If you must use them, do so only with plenty of ventilation.*

Other possibilities might meet your family's needs depending on the severity of the cold and the resources available around your home. A camping family might have a catalytic heater (a gas or oil-fueled heater which provides heat with no flame.) One of these units can keep one room livable in cold weather. A travel trailer or camper can be inhabited in the winter if it has a heater. More than one farm family has been known to take refuge in the relative warmth of a livestock barn under extreme conditions. For shorter periods, there is the family car, a last resort which will be dangerous without proper ventilation.

Bed may be the safest, warmest place for short periods. Use of adequate blankets and coverings will trap and conserve vital body heat, and two or more people in the same bed can share heat. This is an especially good way to keep children warm.

Don't overlook the possibility of solar heat. An appreciable amount of heat can be gained through large windows on the southern side of the house. Sunlight can give a good boost to morale, too!

Providing Fuel for Heating

Many combustibles can be considered for fuel. Some of the common ones include:

Furnace coal	Newspapers, magazines
Cannel coal	Charcoal fluid lighter
Stove coal	Kerosene, gasoline
Furnace oil	Firewood and scraps
Wood chips	Motor oil
Campstove fuel	Fats, grease
Charcoal briquets	Corncobs

1. Coal can be burned in a fireplace or stove if a grate is fashioned to hold it, allowing air to circulate underneath. "Hardware cloth" screening placed on a standard wood grate will keep coal from falling through.
2. Paper "logs" can be made by rolling newspapers or magazines tightly into small log-sized bundles, which can be burned if they are stacked to allow proper air circulation.
3. There may be plenty of burnable wood around, including lumber and furniture if the situation is critical.

Store fuels you will be using in a handy place, but not in the heated area. This is particularly true for highly combustible items such as gasoline, kerosene, and papers.

Which Room Should be Heated?
The decision may be dictated by location of a fireplace, stove, or chimney flue. Be guided by the following:

1. Confine emergency heat to a small area.
2. Try to select a room on the "warm" side of the house away from prevailing cold winds. Avoid rooms with large windows or uninsulated walls. Interior bathrooms probably have the lowest air leakage and heat loss. Your basement may be a good place in cold weather because of the heat gain from the earth.
3. Isolate the room from the rest of the house by keeping doors closed, hanging bedding or heavy drapes over entryways or erecting temporary partitions of cardboard or plywood.
4. Hang drapes, bedding, shower curtains, etc., over doors and windows, especially at night.

Check Your Efforts for Safety

As soon as emergency heating is working and the room protected against exterior cold, stop and appraise the safety of your situation. If there is any hazard or question of safety posed by the emergency heating, make changes immediately. Check carefully for fire hazards.

All heaters except electric heaters should be vented to provide oxygen for complete combustion and to safely remove exhaust gases and smoke.

DO NOT ATTEMPT TO BURN ANYTHING LARGER THAN CANDLES IN YOUR HOME WITHOUT PROVIDING ADEQUATE VENTILATION TO THE OUTSIDE.

Asphyxiation from lack of oxygen or poisonous gases is a great danger when there is not enough ventilation. There is no simple rule for determining how much ventilation you need. For safety, provide cross ventilation by opening a window an inch on each side of a room. It is better to let in some cold air than to take a chance of not having enough air.

As an additional safety factor you should have a firewatch whenever emergency heat is being used. One person should stay awake to watch for fire and to detect the possibility of inadequate ventilation. Drowsiness is one sign of carbon monoxide poisoning. If the firewatch feels sleepy, it may be a sign of not enough ventilation.

Section 8

PASSIVE ENERGY DESIGN

An Energy Conscious Home is one which goes beyond conventional energy conserving features such as insulation in the right places, double glazing and weatherstripping at all openings. It incorporates passive design ideas and/or solar energy systems in its planning, design, construction and use.

This section introduces you to some of these ideas and approaches—approaches which not only produce additional energy savings, but which also make use of solar energy in a passive way. Whether you presently own a home, or whether you plan to build one, you will find many of the ideas presented here of value.

What Do We Mean By "Passive Design Ideas"?

Passive design ideas or approaches,
— use solar energy naturally,
— contain little mechanical hardware,
— require little or no energy themselves, and
— tend to be low in cost.

Sometimes it's easiest to understand passive ideas by contrasting them with "active" design examples. Furnaces, boilers, electric water heaters and air conditioners all fall into the active area: they require complex, expensive and energy-consumer equipment. An active approach to solar heating and cooling uses a carefully-designed, complex and sophisticated solar collector, with fans, pumps, storage or heat exchange units and sophisticated controls. In contrast, one passive approach to solar heating or cooling is a regular window, of the right size, with the right orientation to the sun, designed to capture natural breezes, with insulated window shutter and sufficient heat storage mass.

What Is An Energy Conscious Home?

An energy conscious home approaches the conservation of energy through passive design ideas, and through the use of solar energy, in its planning, design, construction and use. Location configuration, orientation to the sun and breezes, layout, method of construction, and particular design details are all carefully considered. Like the better energy conserving homes now being constructed and renovated around the country, the energy conscious home is well insulated, and incorporates double glazing and weatherstripping where they are needed—but it goes well beyond that. And its energy savings go well beyond those found in conventional homes, too.

Using This Section

This section sketches a number of passive design approaches and ideas. For each approach, it shows one or two ways of incorporating it into homes (there are many more approaches which can be taken), and in most cases it indicates a possible ENERGY SAVINGS (expressed as a percentage reduction).

The energy savings figure is an important one. It is derived by comparing the heating energy used in the design shown with a more conventional design of a 1,600 square foot "standard practice home" located in a region with a fairly cold climate, such as that found in upstate New York. The standard practice home already incorporates good current practice (insulation, weatherstripping, double glazing, 65° winter thermostat settings).

Making The Decision

The passive design ideas in this are a diverse group, and many may not apply to your situation:
— Some make more sense in cold climates, and others in warmer climates.

— Some can be incorporated into existing homes—these are clearly marked with an asterisk—and others are applicable only in new construction.
— Some may appeal to you, and others may not.

It is up to you to select those passive design ideas which you may wish to consider in your home. Hopefully, this raises your own energy consciousness and stimulates you to consider some of the ideas shown. The booklet, though, provides only quick sketches; you may want to consult with architects, builders or developers who have the technical resources to do a more complete job of determining the full impact of these ideas in YOUR situation.

Some of the passive design ideas presented here have a significant impact on the design of the house. The examples presented here are only illustrative. If you are intrigued with the idea, please don't consider the design solution shown here the "only way to go."

We suggest that, once you have pinpointed those approaches which are of interest to you, you consult a number of architects or builders. They are in a position to discuss the specific implications of these ideas for you and your situation—including specific energy savings and impacts on building cost.

It Begins With the Site

Energy conscious design begins with choosing a site which offers opportunities to conserve energy as well as to capture energy from the sun and cooling breezes for heating and cooling your home. Here are some suggestions.

To capture solar energy, face your main living areas to the south and provide sufficient glass area to allow the solar radiation to enter.

Once the solar radiation enters it has to be stored, retained or prevented from being lost through the walls. There are numerous ways of storing and retaining this solar energy. The building's walls, floors, and ceilings

South-Facing Glass Facade

can act as storage devices, as can furniture and special heat storage componants.

You can use prevailing summer breezes to cool your home. The ideal orientation of the side of the house through which the breezes should enter is an oblique angle of 20° to 70° between the wall and wind direction. This will maximize the natural ventilation in the interior. Call your local weather bureau to obtain the prevailing wind direction in your area.

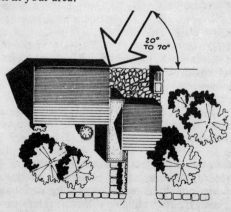

However, locating the house is always a compromise between south facing windows for heating in the winter, and capturing the breezes for cooling in the summer. To protect your facades from winter winds, locate evergreens, fences, and earth berms on the north and west side of your home.

If you are located on a hill and or near a lake, the following additional rules are relevant:

Near a body of water, breezes move from the water to the land during the day, and flow in reverse at night.
On a south-facing hill, breezes tend to move up the hill during the day, and downhill at night.

No matter how your home is sited, you can increase natural ventilation and cooling by using casement-type windows, or partially-opened shutters, on the windward side of the building. These projections create mini-pressure zones in front of the window openings, and increase the velocity of the breeze passing into the openings.

General Rules

When designing with these concepts in mind, there are some general guidelines that you may use in making appropriate decisions. How you use the site relates to several other factors.

SOUTH-FACING GLASS. To capture the necessary solar radiation, it is necessary to provide a minimum amount of south-facing glass. The minimal: 1/4 to 1/5 of the floor area (in temperate climates); 1/3 to 1/4 of the floor area (in colder climates). Combined with proper heat storage mass and insulating shutters, these glass areas can provide 50% or more of the building's space heating needs in warmer climates.

HEAT STORAGE. Heat may be stored in containers filled with water, masonry walls and floors, flower boxes, rock or sand beds, and even furniture. The storage capacity required depends on the amount of radiation captured and the building's use characteristics. In temperate climates, for example, it is necessary to provide 30 pounds of water or 150 pounds of rock storage for each square foot of south-facing glass. If the storage medium cannot be directly exposed to the sun, this number will have to be increased by as much as four times. The ratios of floor area and heat storage surface area should be a minimum of 1:1 to gain the maximum benefit of passive solar heating.

SHADING. To prevent excess heat gain in the summer, provide window shading. These devices (overhangs, grilles, awnings, etc.) should shade the total glass area at noon during the hottest months. Careful attention to orientation and the sun's path through the sky will be required during design. For additional shading, use deciduous trees on the south side of the house; they block the summer sun while allowing the winter sun to penetrate.

SHUTTERS. To prevent heat loss at night, insulated shutters should be closed over the glass areas. During the summer the process can be reversed to gain some cooling: close the shutters during the day, open them at night, exposing the glass areas to the cool night.

What Type Of House Saves Energy?

People have different tastes in housing: some of us like split ranch, some the two-story colonial, while others prefer more contemporary design. Irrespective of style, however, the shape and overall volume of your house can contribute to savings—or excesses—when it comes to energy consumption. Shape and volume can also maximize the use of solar energy through passive means.

To provide some basis for comparison, let's use a basic STANDARD PRACTICE HOUSE which is not unlike many single-family homes being built and occupied around the country today.

The Standard Practice House has 1,600 square feet of living space on two floors, as well as a full basement. It already includes conventional energy-conserving features such as full insulation, storm windows and weatherstripping.

To illustrate the energy savings possible for each passive design idea, we first modify the Standard Practice House to incorporate it—WITHOUT changing its floor area or complement of rooms. Then we calculate the amount of heating—and sometimes cooling—energy required. Comparing the energy required in the modified Standard Practice

House to that required before the modification was introduced gives us a percentage energy saving, and this figure is included with each passive design idea. All calculations are based on the assumption that the house is located in a cold region such as upstate New York.

One Story Rectangular House   4%

A large amount of the total heat loss in a home during the winter occurs through exterior walls—and particularly through and around windows and doors in those walls. Reconfiguring the Standard Practice House into a one-story rectangle does several things:

The total exterior wall area is reduced.

The interior rooms have less exposure to the outside, and window area is also reduced somewhat.

Roof area is increased. In conventional construction it is possible to include more insulation in the roof than in the walls, thus further reducing heat loss. This is not to suggest that all one-story homes are more energy conscious than two-story homes. Trading wall area for roof area, however, can save heating energy.

If reducing wall area can save energy, than it follows that it makes sense to minimize the PERIMETER of the house. Consider the next two approaches.

Circular Floor Plan   9%

The smallest perimeter for a given area is a circle; this in turn gives us the minimum exterior surface area and the savings indicated.

Some, however, do not like round houses or find it awkward to accomodate their life styles in them. The next

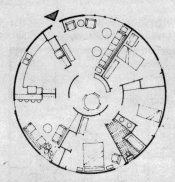

best alternative is the square floor plan: it saves energy, but not as much as the round plan.

One Story Square Floor Plan   5%

Windows can contribute between 15% to 35% of the total heating energy lost in a house; so placing them strategically can make a significant difference in both reducing heatloss and maximizing solar gain. To illustrate this point, let's stay with our square configuration floor plan and see what happens if we take the windows and face them into an atrium which is covered with skylight.

Square Plan With Atrium   21%

This plan includes the same floor and window area as the original Standard Practice House, but the windows are exposed to the inner atrium instead of to the exterior.

The atrium is unheated, but because of the skylight, its winter temperature is not as low as that outside.

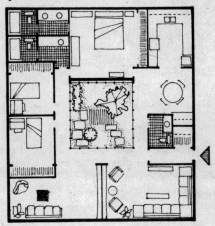

The energy savings shown results only from the reduction in heat loss through the windows and walls. The savings can be doubled or tripled if the atrium is also used (with insulated shutters) as a passive solar collector. The atrium design shown can also be used to reduce heat gain in warm weather if proper sun control and shading devices are used.

The important lesson from these examples is that minimum surface and window areas can save large amounts of energy. Further, locating windows for passive solar collection can significantly increase those savings.

Using Earth To Save Energy

Earth, like most other materials, does provide some thermal resistance, but in principle earth is not a good insulator. The benefit of earth is derived from its capacity to moderate temperature change. Earth slows down the temperature variations between interior and exterior, and it provides protection from cold winter winds. Thus it can contribute a great deal to reducing a building's heat loss.

One important thing to keep in mind is that with earth you must use insulation—in this case placed between the earth cover and the exterior of the structure. This arrangement gives the walls, roofs and ceilings the heat-storage capacity previously described.

To illustrate the impact of working with earth, let us go back to our Standard Practice House. There, all the bedrooms were on the second floor; in the illustration below, they have been relocated below the first floor in what used to be the basement.

Bedrooms Below Finished Grade   23%

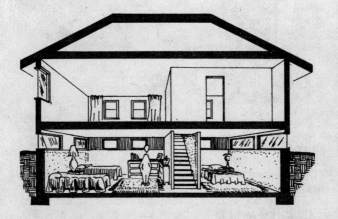

In doing this, the amount of excavation is not increased, but the first floor is raised slightly to maintain a 7'-6" ceiling height and to accommodate 2' high clerestory windows in the "under-the-ground" floor level. This makes the space much more liveable and saves considerable energy since much of the exterior wall is exposed to earth rather than to the outside. In the summertime, or in warm climates, this approach also reduces heat gain, requiring less cooling energy.

Another approach to using earth is to berm it up against the walls of the house. When conventional windows, say with 3' high sills, are used, earth can be bermed to the first-floor sills for both the Standard Practice House and the Square Plan House.

Earth Berming to Window-Sill
Square Plan   **13%**

Earth Berming to Roof Eave   **32%**

If clerestory windows can be used, and earth bermed to their sills, additional savings are possible. The temperature 4' to 5' below grade is relatively constant at around 55°F, and a duct located in the berm, with a small fan, provides a simple passive cooling system!

It should be noted that earth-berming and the introduction of below-grade spaces in houses requires careful attention to waterproofing, foundation drainage, means of egress and humidity control. Further, earth berming existing buildings requires special design and technical expertise: treatment of existing surfaces for protection from moisture, rodents, insects and even tree roots is required. Consult an expert for these applications.

The prospect of energy savings in underground houses is now becoming well known. This approach not only saves heating energy but also offers an excellent way to keep cool without an air conditioning system in warmer climates.

Some Other Exterior Features

Having illustrated how plan, configuration, window placement and earth can contribute to energy savings through passive design, let's move on to some other exterior features which can produce energy savings.

Entry Lock Added to
Standard Practice House   **5%**

Doors are important penetrations in exterior walls, and where they open directly into the interior of the house, large amounts of heated or cooled air may escape each time they are opened. An entry lock, either designed into the interior of the house or added to the exterior, reduces energy loss by providing two doors (only one of which is normally open at any moment) separated by a small unheated or uncooled air space.

| The | heat-loss . . . . | 8% |
| Greenhouse | heat-gain . . . . | 7% |

Additional savings can be obtained by turning an entry lock into a greenhouse. The "greenhouse effect" is well known: solar radiation through large glass areas will keep temperatures at reasonable levels (even without supplementary heating) in the wintertime; by adding plants and other insulating/shading devices, the effects of high heat gain in warmer weather can be mitigated.

Adding a greenhouse to a home increases the thermal resistance of the outside envelope in two ways:
— the outside temperature of the main exterior wall is increased in cold weather and decreased in warm weather;
— infiltration losses around doors and windows are reduced because the main wall no longer is directly exposed to the elements.

The energy savings shown result only from the increased thermal resistance of the envelope; using a greenhouse as a passive solar collector provides additional savings if it is oriented properly and if heat storage capability is included in its design.

The Bead-Window   34%

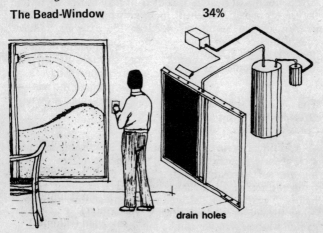

drain holes

One of the advantages of a greenhouse is that it allows the seasonal use of plants and shades to control heat transfer. A more sophisticated approach, based on the patented "bead wall" concept looks at the window as a device with varying thermal resistance—depending on the need. The bead-window consists of two glass or plastic sheets separated by a 3″ air space which can be filled with high thermal resistance styrofoam beads using a pump and blower system; them emptied of the beads using a vacuum system. The calculation assumes that the north windows of the Standard Practice House are eliminated and that bead windows are placed on the remaining three walls of the Standard Practice House. The windows are assumed to be "transparent" to heat gain during these hours:

East windows, transparent from 8 am to 11 am
South windows, transparent from 8 am to 4 pm
West windows, transparent from 1 pm to 4 pm.

| Window | heat-loss . . . . | 28% |
| Shutters | heat-gain . . . . | 34% |

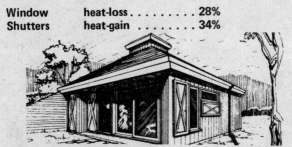

A more conventional approach to "adjusting" the thermal resistance of windows is to use shutters—IF they have genuine insulating value. The shutters shown have a wood face and an insulating core; once applied to the windows of the Standard Practice House, the savings shown result if they are opened on the following schedule:

East wall, opened from 8 am-11 am (winter)
1 pm-4 pm (summer)
South wall, opened from 8 am-4 pm (winter)
7-9 am; 3-5 pm (summer)
West wall, opened from 1 pm-4 pm (winter)
8 am-11 am (summer)
North wall, opened from any three hours (winter)
8 am-7 pm (summer)

Speaking of window shutters, it is possible to conceive of a window shutter which, in its open position, functions as a solar collector with a self-contained storage unit. When closed, the shutter/collector vents the stored heat directly into the room.

Solar Window room savings **54%**
Shutter   house savings. **6%**

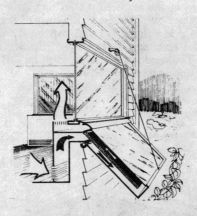

The hinged solar window shutter can be inclined to respond to the appropriate radiation angle and it can be closed against the window frame. The shutter retains the heat generated by solar radiation, and when closed, operable vents facilitate heat transfer into the room.

The calculation assumes the use of such shutters in one of the bedrooms of the Standard Practice House where one window faces east and another faces south. The second figure given indicates the impact of our example on the total energy consumption of the house.

Another approach to passive solar systems is to integrate a solar collector with the window assembly as shown. The collector incorporates collection, storage and direct venting into the interior room. To produce the savings indicated, the collector-windows are used in the southeast bedroom of the Standard Practice House.

Solar Window Unit   **62%**

Looking at solar collectors, it's possible to take passive design approaches to incorporating them into the home.

One approach is to integrate collectors into earth berming as shown. The berm angle should be the same as the average solar radiation angle for the locale. The collector unit is self-contained, and heat is transferred into the building by convection and by control of manually-operated vents.

Solar Collectors On Earth Berms

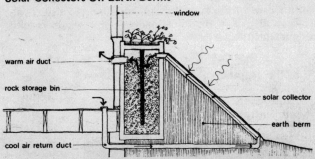

window

warm air duct

rock storage bin

solar collector

earth berm

cool air return duct

Earlier the importance of south-facing glass in capturing solar radiation was noted. In the illustrations shown here, the entire south-facing facade is treated as a "solar collector wall."

The first drawing depicts the principles behind the solar collector wall. Sun penetrating through the glass strikes the blackened surface of the masonry or concrete wall—simultaneously heating both the wall and the air space between the glass and the wall. The heated air in the space rises and enters the room at the top. Cool air enters at the bottom of the wall, to be heated in a continuous cycle (the principle is called "thermosyphoning"). In addition, the heat in the masonry wall migrates to the inside and, when the sun disappears, the wall acts as a radiator. Careful use of insulation allows use of the wall for cooling at night in the summer, too.

Vertical Solar Collector and Heat Storage Wall

Solar Collector Wall   **27%**

The second drawing illustrates the conversion of the south facade of the Standard Practice Home to a solar collector wall. All of the doors and operable windows have been retained, however.

The Drum Wall

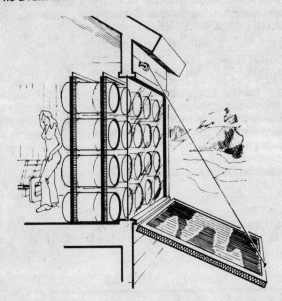

In a variation of the solar collector wall, the "drum wall" includes large drums of water placed in racks behind the window glass area. The exterior of the drum is painted black and the solar energy coming through the window is collected in the drums which, in turn, act as radiators to heat the room. At night large window shutters are closed over the outside of the glass area to prevent heat loss.

What Happens When You Put Some Of These Ideas Together?

So far we have dealt with a number of individual ideas—planning and design features which can reduce the amount of energy consumed in the home. What happens if you incorporate several of these features in your home?

As was pointed out earlier, energy savings from the passive design ideas shown are not necessarily additive. Each house presents its own situation—and requires its own energy consumption calculations.

To provide an illustration, let's take a number of the passive energy saving ideas and incorporate them into an ENERGY CONSCIOUS HOUSE.

Energy	heat-loss	32%
Conscious House	heat-gain	23%
Floor Plan	hot water	36%

The Energy Conscious House contains 1,600 square feet (excluding the atrium) and is analyzed under the same conditions as the earlier Standard Practice House. From the illustrations, you can see that the following features have been used:

One-story Configuration
Minimum Perimeter Distance
Window Shutters
Atrium and Entry Locks
Earth Berming
Maximum Insulation in Roof and Walls
Weatherstripping and Storm Windows

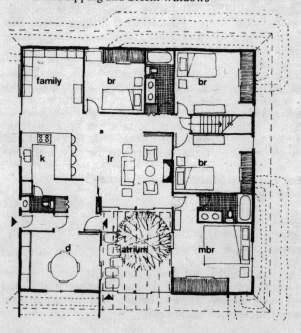

Summing Up

Energy conscious design can provide significant energy savings in homes. It begins with choosing the right site and property locating the house to take advantage of the sun and wind as well as other natural forces. The house's configuration plan, exterior features and interior characteristics all contribute.

Most of the passive design ideas presented here are ready-to-implement; a few now exist only as concepts or prototypes. All can be considered in looking at new construction; many can also be incorporated into existing homes. Some, but not all, require special professional or technical expertise.

Which passive design approaches to consider? How much energy can you save? What will it cost?

There are no simple answers to these questions. Each home is unique. Your own needs, desires and priorities are critical factors in looking at your housing. Further, climatic conditions and construction approaches vary from place to place.

To optimize YOUR situation, it is suggested that you seek the assistance of an energy-conscious architect and/or builder early in the game. You can discuss your needs, your likes and dislikes, and your available resources. In return, you can gain information on expected costs and savings.

Energy consciousness now may reap significant—and continuing—dividends as you occupy your home year after year.

Section 9

SOLAR HOT WATER SYSTEMS

Functional Description

The basic function of a solar domestic hot water system is the collection and conversion of solar radiation into usable energy. This is accomplished—in general terms—in the following manner: Solar radiation is absorbed by a *collector*, placed in storage as required, with or without the use of a *transport* medium, and *distributed* to point of use. The performance of each operation is maintained by automatic or manual *controls*. An *auxiliary energy system* is usually available both to supplement the output provided by the solar system and to provide for the total energy demand should the solar system become inoperable.

The parts of a solar system—collector, storage, distribution, transport, controls and auxiliary energy—may vary widely in design, operation, and performance. They may be arranged in numerous combinations dependent on function, component compatibility, climatic conditions, required performance, site characteristics, and architectural requirements.

Of the numerous concepts presently being developed for the collection of solar radiation, the relatively simple flat-plate *collector* has the widest application. It consists of an absorber plate, usually made of metal and coated black to increase absorption of the sun's energy. The plate is insulated on its underside and covered with one or more transparent cover plates to trap heat within the collector and reduce convective losses from the absorber plate. The captured heat is removed from the absorber plate by means of a heat transfer fluid, generally air or water. The fluid is heated as it passes through or near the absorber plate and then transported to points of use, or to storage, depending on energy demand. (Most solar hot water systems use liquid heat transfer fluids.)

The *storage* of thermal energy is the second item of importance since there will be an energy demand during the evening, or on sunless days when solar collection cannot occur. Heat is stored when the energy delivered by the sun and captured by the collector exceeds the demand at the point of use. In some cases, it is necessary to transfer heat from the collector to storage by means of a heat exchanger. In other cases, transfer is made by direct contact of the heat transfer fluid with the storage medium.

The *distribution* component receives heat energy from the collector or storage, and dispenses it at points of use as hot water.

The *controls* of a solar system perform the sensing, evaluation and response functions required to operate the system in the desired mode. For example, when the collector temperature is sufficiently higher than storage temperature, the controls will cause the heat transfer fluid in storage to circulate through the collector and accumulate solar heat.

An *auxiliary energy* system provides the supply of energy when stored energy is depleted due to severe weather or clouds. The auxiliary system, using conventional fuels such as oil, gas, electricity, or wood provides the required heat until solar energy is available again.

Most solar systems can be characterized as either active or passive in their operation.

An *active solar system* is generally classified as one in which an energy resource—in addition to solar—is used for the transfer of thermal energy. This additional energy, generated on or off the site, is required for pumps or other heat transfer medium moving devices for system operation.

Generally, the collection, storage, and distribution of thermal energy is achieved by moving a transfer medium throughout the system with the assistance of pumping power.

A *passive solar system,* on the other hand, is generally classified as one where solar energy alone is used for the transfer of thermal energy. Energy other than solar is not required for pumps or other heat transfer medium moving devices for system operation. Collection, storage, and distribution is achieved by natural heat transfer phenomena employing convection, radiation and conduction.

A Residence with Solar Collectors Incorporated Into the Roof.

Operational Description

The solar hot water system usually is designed to preheat water from the incoming water supply prior to passage through a conventional water heater. The domestic hot water preheat system can be combined with a solar heating system or designed as a separate system. Both situations are illustrated.

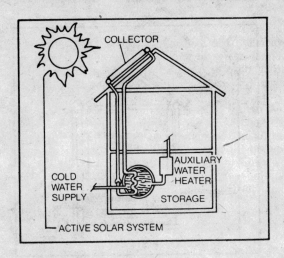

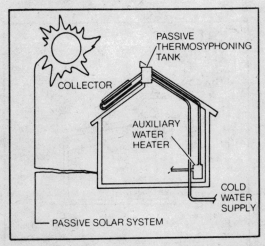

Domestic Hot Water Preheating—Separate System. Domestic hot water preheating may be the only solar system included in many designs. An active solar system is shown in the upper figure and a passive thermo-syphoning arrangement in the lower.

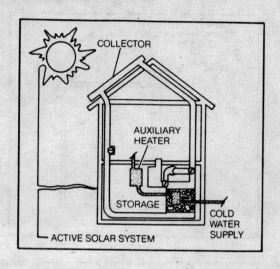

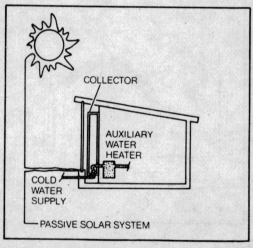

Domestic Hot Water Preheating—Combined System. Domestic hot water is preheated as it passes through heat storage enroute to the conventional water heater. An active solar system using air for heat transport is shown in the upper figure and a passive solar system in the lower.

Basics of Solar Energy Utilization

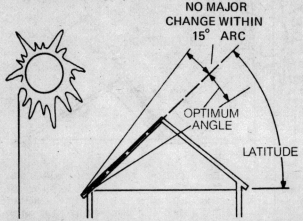

Collector Tilt for Domestic Hot Water. The optimum collector tilt for domestic water heating alone is usually equal to the site latitude. Tilt angle for solar hot water can be latitude ± 15 degrees for optimum performance; a 15 degree variation does not cause a major impact.

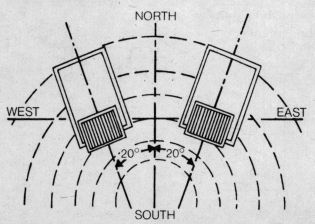

Collector Orientation. A collector orientation of 20 degrees to either side of true South is acceptable. However, local climate and collector type may influence the choice between East or West deviations.

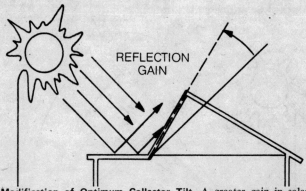

Modification of Optimum Collector Tilt. A greater gain in solar radiation collection sometimes may be achieved by tilting the collector away from the optimum in order to capture radiation reflected from adjacent ground or building surfaces. The corresponding reduction of radiation directly striking the collector, due to non-optimum tilt, should be recognized when considering this option.

Snowfall Consideration. The snowfall characteristics of an area may influence the appropriateness of these optimum collector tilts. Snow buildup on the collector, or drifting in front of the collector, should be avoided.

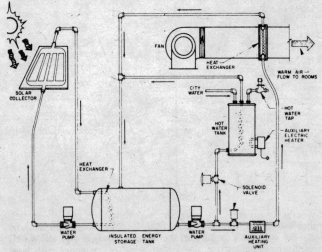

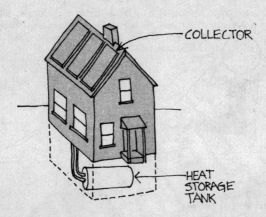

COLLECTOR

HEAT
STORAGE
TANK

Shading of Collector

Another issue related to both collector orientation and tilt is shading. Solar collectors should be located on the building or site so that unwanted shading of the collectors by adjacent structures, landscaping or building elements does not occur. In addition, considerations for avoiding shading of the collector by other collectors should also be made. Collector shading by elements surrounding the site must also be addressed.

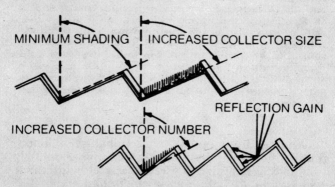

MINIMUM SHADING   INCREASED COLLECTOR SIZE

INCREASED COLLECTOR NUMBER

REFLECTION GAIN

Self-Shading of Collector. Avoiding all self-shading for a bank of parallel collectors during useful collection hours (9 AM and 3 PM) results in designing for the lowest angle of incidence with large spaces between collectors. It may be desirable therefore to allow some self-shading at the end of solar collection hours, in order to increase collector size or to design a closer spacing of collectors, thus increasing solar collection area.

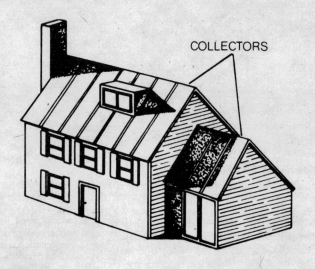

COLLECTORS

Shading of Collector by Building Elements. Chimneys, parapets, fire walls, dormers, and other building elements can cast shadows on adjacent roof-mounted solar collectors, as well as on vertical wall collectors. The drawing above shows a house with a 45° southfacing collector at latitude 40° North. By mid-afternoon portions of the collector are shaded by the chimney, dormer, and the offset between the collector on the garage. Careful attention to the placement of building elements and to floor plan arrangement is required to assure that unwanted collector shading does not occur.

Active Systems

Active solar systems are characterized by collectors, thermal storage units and transfer media, in an assembly which requires additional mechanical energy to convert and transfer the solar energy into thermal energy. The following discussion of active solar systems serves as an introduction to a range of active concepts which have been constructed.

Domestic hot water can be preheated either by circulating the potable water supply itself through the collector, or by passing the supply line through storage enroute to a conventional water heater. Three storage-related preheat systems are shown below.

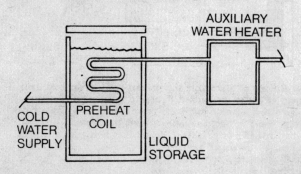

Preheat Coil in Storage. Water is passed through a suitably sized coil placed in storage enroute to the conventional water heater. Unless the preheat coil has a protective double-wall construction, this method can only be used for solar systems employing non-toxic storage media.

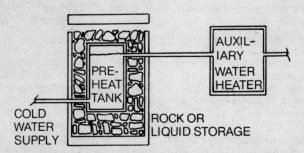

Preheat Tank in Storage. In this system, the domestic hot water preheat tank is located within the heat storage. The water supply passes through storage to the preheat tank where it is heated and stored, and later piped to a conventional water heater as needed. A protective double wall construction again will be necessary unless a non-toxic storage medium is used.

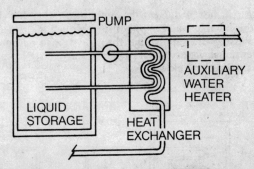

Preheat Outside of Storage. In this preheat method, the heat transfer liquid in storage is pumped through a separate heat exchanger to be used for domestic hot water preheating. This separate heat exchanger could be the conventional water heater itself. However, if the liquid from storage is toxic, the required separation of liquids is achieved by the use of a double-wall exchanger, as diagrammed, in which the water supply simply passes through enroute to the conventional water heater.

Collector Mounting

Flat-plate collectors are generally mounted on the ground or on a building in a fixed position at prescribed angles of solar exposure—angles which vary according to the geographic location, collector type, and the use of the absorbed heat. Flat-plate collectors may be mounted in four general ways as illustrated below.

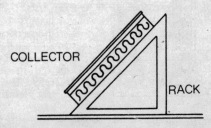

Rack Mounting. Collectors can be mounted at the prescribed angle on a structural frame located on the ground or attached to the building. The structural connection between the collector and the frame and the frame and the building or site must be adequate to resist any impact loads such as wind.

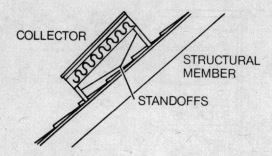

COLLECTOR

STRUCTURAL
MEMBER

STANDOFFS

Stand-Off Mounting. Elements that separate the collector from the finished roof surface are known as stand-offs. They allow air and rain water to pass under the collector thus minimizing problems of mildew and leakage. The stand-offs must also have adequate structural properties. Stand-offs are often used to support collectors at an angle other than that of the roof to optimize collector tilt.

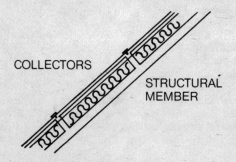

COLLECTORS

STRUCTURAL
MEMBER

Direct Mounting. Collectors can be mounted directly on the roof surface. Generally, the collectors are placed on a water-proof membrane on top of the roof sheathing. The finished roof surface, together with the necessary collector structural attachments and flashing, are then built up around the collector. A weatherproof seal between the collector and the roof must be maintained, or leakage, mildew, and rotting may occur.

Integral Mounting. Unlike the previous three component collectors which can be applied or mounted separately, integral mounting places the collector within the roof construction itself. Thus, the collector is attached to and supported by the structural framing members. In addition, the top of the collector serves as the finished roof surface. Weather tightness is again crucial to avoid problems

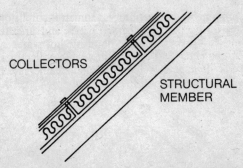

COLLECTORS

STRUCTURAL MEMBER

of water damage and mildew. This method of mounting is frequently used for site built collectors.

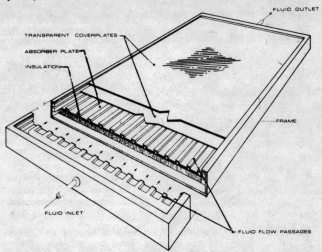

FLUID OUTLET

TRANSPARENT COVERPLATES

ABSORBER PLATE

INSULATION

FRAME

FLUID INLET

FLUID FLOW PASSAGES

Detailed view of a typical solar collector.

Component Description

As noted in the functional description, a solar domestic hot water system is composed of numerous individual parts and pieces including: collectors; storage; a distribution network with pipes, pumps and valves; insulation; a system of manual or automatic controls; and possibly heat exchangers, expansion tanks and filters. These parts are assembled in a variety of combinations depending on function, component compatibility, climatic conditions, required performance, site characteristics and architectural requirements, to form a solar domestic hot water system. Some components that are unique to the collector system or that are used in an unconventional manner are briefly illustrated and discussed in the next few pages.

Flat-Plate Collectors

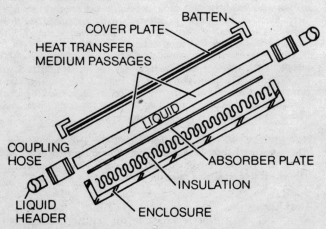

The flat-plate collector is a common solar collection device used for domestic water heating. Most collectors are designed to use liquid (usually treated water) as the heat transfer medium. However, air systems are available for domestic water heating. Most flat-plate collectors consist of the same general components, as illustrated.

BATTENS serve to hold down the cover plate(s) and provide a weather tight seal between the enclosure and the cover.

The COVER PLATE usually consists of one or more layers of glass or plastic film or combinations thereof. The cover plate is separated from the absorber plate to reduce reradiation and to create an air space, which traps heat by reducing convective losses. This space between the cover and absorber can be evacuated to further reduce convective losses.

TUBES are attached above, below or integral with an absorber plate for the purpose of transferring thermal energy from the absorber plate to a heat transfer medium. The largest variation in flat-plate collector design occurs with this component and its combination with the absorber plate. Tube on plate, integral tubes and sheet, open channel flow, corrugated sheets, deformed sheets, extruded sheets and finned tubes are some of the techniques used.

Since the ABSORBER PLATE must have a good thermal bond with the fluid passages, an absorber plate integral with the heat transfer media passages is common. The absorber plate is usually metallic, and normally treated with a surface coating which improves absorptivity. Black or dark paints or selective coatings are used for this purpose. The design of this passage and plate combination helps determine a solar system's effectiveness.

INSULATION is employed to reduce heat loss through the back of the collector. The insulation must be suitable for the high temperature that may occur under no-flow or dry-plate conditions, or even normal collection operation. Thermal decomposition and outgassing of the insulation must be prevented.

The ENCLOSURE is a container for all the above components. The assembly is usually weatherproof. Preventing dust, wind and water from coming in contact with the cover plate and insulation is essential to maintaining collector performance.

Heat Exchangers

A heat exchanger is a device for transferring thermal energy from one fluid to another. In some solar systems, a heat exchanger may be required between the transfer medium circulated through the collector and the storage medium or between the storage and the distribution medium. Three types of heat exchangers that are most commonly used for these purposes are illustrated below.

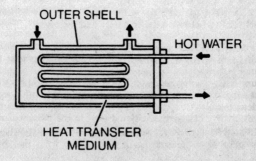

Shell and Tube. This type of heat exchanger is used to transfer heat from a circulating transfer medium to another medium used in storage or in distribution. Shell and tube heat exchangers consist of an outer casing or shell surrounding a bundle of tubes. The water to be heated is normally circulated in the tubes and the hot liquid is circulated in the shell. Tubes are usually metal such as steel, copper or stainless steel. A single shell and tube heat exchanger cannot be used for heat transfer from a toxic liquid to potable water because double separation is not provided and the toxic liquid may enter the potable water supply in a case of tube failure.

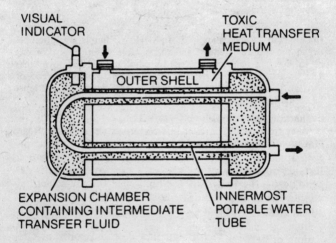

VISUAL
INDICATOR

TOXIC
HEAT TRANSFER
MEDIUM

OUTER SHELL

EXPANSION CHAMBER
CONTAINING INTERMEDIATE
TRANSFER FLUID

INNERMOST
POTABLE WATER
TUBE

Shell and Double Tube. This type of heat exchanger is similar to the previous one except that a secondary chamber is located within the shell to surround the potable water tube. The heated toxic liquid then circulates inside the shell but around this second tube. An intermediary non-toxic heat transfer liquid is then located between the two tube circuits. As the toxic heat transfer medium circulates through the shell, the intermediary liquid is heated, which in turn heats the potable water supply circulating through the innermost tube. This heat exchanger can be equipped with a sight glass to detect leaks by a change in color—toxic liquid often contains a dye—or by a change in the liquid level in the intermediary chamber, which would indicate a failure in either the outer shell or intermediary tube lining.

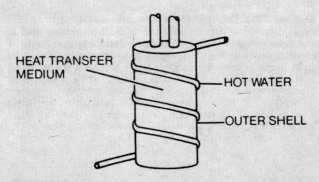

HEAT TRANSFER MEDIUM

HOT WATER

OUTER SHELL

Double Wall. Another method of providing a double separation between the transfer medium and the potable water supply consists of tubing or a plate coil wrapped around and bonded to a tank. The potable water is heated as it circulates through the coil or through the tank. When this method is used, the tubing coil must be adequately insulated to reduce heat losses.

Evaluating Solar Hot Water Systems

A typical solar domestic hot water system for an average family of four may cost $2,200 uninstalled, depending on geographic location, collector efficiency and other factors. Installation costs vary greatly on a case by case basis depending on the design of the home and on any structural modifications required. The system should be HUD Program approved; contact your State Energy Office.

Two important factors are first cost and collector efficiency. Both factors must be considered when comparing collectors. Thus a relatively inexpensive collector with a low efficiency may be a poor choice when compared to a more expensive one that captures and delivers the sun's energy more efficiently. All other things being equal, the collector that delivers more *heat per dollar* should be selected.

How long is it expected to last? Is it weatherproof and does it shed water?

If something goes wrong, who will fix it, how long will it take to fix it, and are repair parts easily obtainable? And what will various repairs cost?

Is it susceptible to either freezing or overheating? There are a number of adequate solutions to both problems, and the collector must have protection for both extremes built in. In addition, if antifreeze fluid is used in the collector, it must not mix with water for domestic use.

Metal corrosion can cause irreparable damage to a solar system and shorten a system's life span. More importantly, corrosion can cause serious health problems if the water is used directly by the user. The three metals commonly used in collectors are copper, steel and aluminum. Inhibitors are usually added to prevent corrosion in most systems. Copper will function over a long period without inhibitors, while steel and aluminum may fail quickly without special protection. With inhibited water, all three will last indefinitely as long as inhibition is maintained.

The greater the number of joints, the greater the possibility of leakage. Some liquid collectors using channel systems have reduced the need for soldered joints considerably without affecting efficiency. Know what happens if a leak occurs, how it will be fixed, who will fix it and how much it will cost.

Try to obtain the names of other buyers, and learn whether the buyer is satisfied with its performance and whether the collector has lived up to the claims of the seller.

The pipes, ducts, and back parts of collectors, and storage tank should be insulated to prevent heat loss. This is true of inside and outside pipes and ducts. Avoid use of heating tapes to prevent freezing. They may use more energy than the solar system saves.

The flow of water in pipes has to be positive. If the collector is the drain down type, there cannot be any traps in the external pipes where water can collect and possibly freeze. Pipes and collectors need to be pitched to achieve drain down.

Air has a habit of getting into systems even though they are watertight. Facility should be provided for bleeding the system. An expansion tank, or room in storage for expansion, should also be included.

These items use electricity and hence should not be any bigger than necessary to perform. If they are too big then they will reduce the total energy savings. A time elapsed meter can be installed to see if the fans or pumps are running for too long a period. Sometimes they can be on when not needed but since they are quiet the homeowner is not aware that they are on. Pumps should be located below tank water level.

The system should be taken through a complete operating test before acceptance by the owner. All controls should be confirmed as working and all leaks should be

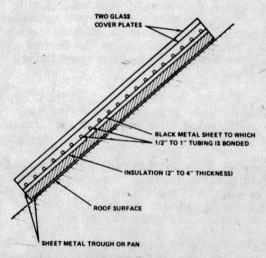

TWO GLASS COVER PLATES

BLACK METAL SHEET TO WHICH 1/2" TO 1" TUBING IS BONDED

INSULATION (2" TO 4" THICKNESS)

ROOF SURFACE

SHEET METAL TROUGH OR PAN

DIMENSIONS: THICKNESS (A DIRECTION) 3 INCHES TO 6 INCHES
LENGTH (B DIRECTION) 4 FEET TO 20 FEET
WIDTH (C DIRECTION) 10 FEET TO 50 FEET
SLOPE DEPENDENT ON LOCATION AND ON
WINTER SUMMER LOAD COMPARISON

Solar heat collector for home heating and cooling

fixed. There should be confirmation that the collector does increase the temperature in storage. If the system is a drain-down type there should be evidence that the system will drain-down properly.

These items should be corrosive resistant and as silent as possible. In a system that connects directly to the potable water supply inhibitors or anti-freeze cannot be used. In this case the controls and valves should be brass.

DIAGRAM OF ACTIVE SYSTEM

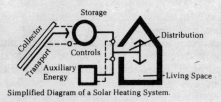

Simplified Diagram of a Solar Heating System.

DIAGRAM OF SOLAR COLLECTOR

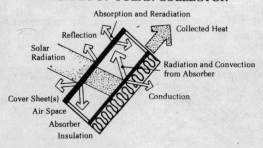

In assessing the economics of any solar energy system for your home, you should look carefully at the trade-offs between the higher installation and equipment costs you will be paying compared to the potential savings you will be realizing due to decreased use of conventional fuel sources. As stated earlier, the more conventional fuel prices rise, the better your solar alternative will look and the economics of solar energy systems will become more attractive.

Section 10

YOUR ENERGY TAX CREDITS

Congress has provided tax incentives to encourage energy conservation and the development of alternate energy sources. This section explains the new tax credits that may be claimed by individuals who installed energy-saving components of solar, geothermal, or wind-powered equipment in their homes after April 19, 1977. For solar, geothermal, or wind-powered equipment, the credit is also available for installations in connection with construction or reconstruction of dwellings before April 20, 1977, provided the original occupancy by the taxpayers began after April 19, 1977.

Residential Conservation

Taxpayers can receive tax credits for installing energy-saving materials in their homes amounting to 15 percent of the first $2,000 spent on qualifying equipment, up to a maximum of $300. The credit can be subtracted from taxes due.

The credit would be for existing dwellings only (in existence as of the effective date of the act). The credit would be effective through Dec. 31, 1985. The installation must be made in the taxpayer's principal residence. Eligible for the credit are owners, renters, and owners of cooperatives or condominiums.

A taxpayer who qualified for a tax credit in excess of the tax he owes could carry the credit forward on future tax returns through the taxable years ending Jan. 1, 1988.

However, for expenditures made in 1977, a credit could only be claimed on the taxpayer's 1978 tax return. No refunds for credit in excess of taxes due would be allowed.

Eligible for the credit would be insulation, furnace replacement burners, devices for modifying flue openings, electrical or mechanical furnace ignition systems that replace a gas pilot light, storm or thermal doors or windows, clock thermostats, caulking or weatherstripping and other items specified by the Treasury Department as increasing energy efficiency in the dwelling. Items include automatic energy-saving thermostats and meters that display the cost of energy usage.

Residential, Solar Wind

A taxpayer could receive up to 40 percent of the first $10,000 spent and 20 percent on the next $8,000 spent on qualifying equipment. The maximum credit would be $4,000. Like the residential credit, it would be available for equipment installed from April 20, 1977, to Dec. 31, 1985, and taxpayers could carry forward any excess credit, except on 1977 expenditures. No refunds would be allowed. Unlike the conservation credit, this credit would apply to both new and existing homes.

Eligible equipment includes solar space and hot water heating, wind energy equipment, geothermal equipment and other "renewable energy" equipment, including "passive solar" energy installations. Passive solar energy relies on the design of the building and the use of thermal storage mass such as masonry walls. In a typical passive system, sunshine enters the structure through a south-facing window, is stored as heat in a stone wall and is radiated back during the night. The Congress adopted language to eliminate from eligibility anything that serves a "significant" structural function, such as extra thick walls. Solar electricity devices, such as photovoltaic cells, leased solar energy equipment and wind transportation equipment would not qualify.

Solar Home Loans

Homeowners at all income levels could receive loans.

at reasonable interest rates for solar heating, cooling, and hot water equipment purchased and installed after the date of enactment. Banks and other community lenders could make solar loans of up to $8,000 each at interest rates ranging from about 7 percent to 12 percent for periods of up to 15 years. Buildings with up to four family dwelling units would be eligible. Another provision of importance to home buyers would increase by up to 20 percent the limits on federally-insured loans; this provision would apply to the loan programs of the Federal Housing Administration and the Farmers Home Administration. For example, an individual interested in buying a solar home with a low-interest FHA loan could exceed the current FHA mortgage limit of $60,000 to buy a solar home valued at $72,000.

The definition of energy-conserving improvements eligible for federal home improvement loan insurance is expanded to include wind energy equipment and active or "passive" solar energy equipment. Passive solar uses building design to make maximum use of solar energy through such means as south-facing windows.

You claim the credits on line 45 of your Form 1040 for 1978. The amount of credit is computed on Form 5695, which you attach to your return. All eligible expenditures from April 20, 1977, through the end of 1978 are included in computing your credit for 1978. For this purpose, expenditures are generally treated as made when the original installation is completed. You may not amend a 1977 return to claim an energy credit for that year. You may not claim an energy credit on Form 1040A.

Fiscal year return. If you file your return on a fiscal year basis, you may not take the credit until your first tax year beginning after 1977. Your eligible expenditures, however, from April 20, 1977, through the end of your fiscal year may be combined in determining your energy credit for the first tax year.

Two distinct energy credits make up the residential energy credit, each with its own conditions and limitations. These credits are based on:

1. Expenditures for home energy conservation; and
2. Expenditures for renewable energy source property.

Residential Energy Credit

The residential energy credit is made up of the *credit for energy conservation expenditures* (15% of the first $2,000 spent on components to conserve energy, or a maximum credit of $300) plus the *credit for renewable energy source expenditures* (30% of the first $2,000 plus 20% of the next $8,000 spent on solar, geothermal, or wind-powered equipment, or a maximum credit of $2,200).

The credit is based on the cost of items installed after April 19, 1977, and before January 1, 1986, regardless of when the items were actually purchased. However, solar, geothermal, or wind-powered items that were installed in a newly constructed or reconstructed residence before April 20, 1977, will qualify for the credit if the residence was first occupied by the taxpayer after April 19, 1977. The cost of qualifying items includes the cost of their original installation.

Your residential energy credit must amount to at least $10 in any one year before it may be claimed. This $10 minimum applies *before* you consider the limitation to tax described in the following paragraph. The minimum applies to either joint or separate returns.

The credit may not exceed the amount of tax for which you are otherwise liable. Specifically, it is limited to the tax on line 37, Form 1040, minus the credits on lines 38 through 44, Form 1040.

Carryover of unused credit. You may carry over an unused credit (that you may not use because it exceeds your tax liability) to the next tax year. An unused credit may continue to be carried over to later tax years through 1987. (For a fiscal year filer, an unused credit may be carried over through the fiscal year beginning in 1987.)

You must reduce the basis of your residence by the amount of residential energy credit allowed, if the items for which the credit is taken are properly added to the basis of the residence.

Both owners and renters of dwellings are eligible for the credit, provided they actually pay for the qualifying items.

Stockholders of cooperative housing corporations and owners of condominium units may claim a credit based on their allocable share of qualifying expenditures made by the cooperative housing corporation or condominium management association for the benefit of the common owners. For a stockholder of a cooperative housing corporation, the allocable share of the cooperative's expenditures is proportionate to the stockholder's share of the cooperative's total outstanding stock.

If you own or rent your residence jointly with others, the overall limits on qualifying expenditures apply to the combined expenditures of all the owners or renters. If the actual amount spent is greater than the limits, the maximum credit must be apportioned to the joint owners or renters based on the portion of the total expenditures that each contributed. The fact that one joint occupant may be unable to claim all or part of the credit, either because of insufficient tax liability or because of not meeting the $10 minimum credit, has no effect on the computation of the credit for the other joint occupants.

Business use of home. If more than 20% of the use of an energy-conserving or renewable energy source item is for business purposes, you must allocate the expenditure for the item between the business use and the residential use. Only the expenditure allocable to the residential use qualifies for the credit. A swimming pool is not considered used for residential purposes.

Home Energy Conservation Expenditures

You are entitled to a credit of 15% of the first $2,000 you spend on components to conserve energy in your home. You are eligible for the credit when the original

installation of the components is completed. The full
$2,000 of energy-saving items need not be installed in a
single tax year. However, if the qualifying items are installed over a period of more than one tax year, the 15%
credit must be claimed for the tax years in which the items
are installed, except that expenditures for items installed
from April 20, 1977, through the end of 1978 must be
combined in computing the 1978 credit. A new $2,000
limit applies to each subsequent principal residence in
which you live.

Example 1. In September 1977 you purchase and install $500 of insulation. In February 1978 you install
storm windows costing $1,500. If you file your tax returns
on a calendar year basis, you may claim a $300 (15% of
$2,000) credit on your 1978 return. You may not amend
your 1977 return to claim a credit for that year.

Example 2. The facts are the same as in Example 1,
but you move to another home in 1979 and spend additional amounts for energy-saving components for the new
home. You are eligible for another credit of up to $300
for these expenditures, provided the other conditions are
met. You may claim the credit for your additional energy-
saving expenditures on the new home, even though the
previous owner of your new home had claimed a credit
for energy-saving expenditures on it.

Principal residence. The dwelling on which you install
the qualified energy-saving components must be your
principal residence, must be located in the United States,
and must have been substantially completed before April
20, 1977. To qualify for the credit, a dwelling is considered your principal residence beginning 30 days before
the date you occupy it.

The energy-saving components you install must be new,
must be expected to last at least 3 years, and must meet
performance and quality standards to be specified by the
Secretary of the Treasury. As of the date of this publication, no performance and quality standards have been
issued. However, components purchased before the per-

formance and quality standards are published need not meet these standards.

Qualifying energy-saving components include the following:

Insulation designed to reduce heat loss or heat gain of a residence or water heater

Storm or thermal windows or doors for the exterior of the dwelling

Caulking or weather stripping of exterior doors or windows

Clock thermostats or other automatic energy-saving setback thermostats

Furnace modifications designed to increase fuel efficiency, including replacement burners, modified flue openings, and ignition systems that replace a gas pilot light

Meters that display the cost of energy usage

Additional items, when specified by the Secretary of the Treasury as increasing the energy efficiency of the dwelling. As of the date of this publication, no additional items have been so specified.

Insulation is any item that is specifically and primarily designed to reduce heat loss or gain of a dwelling or water heater. It includes, but is not limited to, materials made of fiberglass, rock wool, cellulose, styrofoam, urea-based foam, urethane, vermiculite, perlite, polystyrene, reflective insulation, and extruded polystyrene foam. It is installed in one of the following applications:

Ceiling insulation, which is installed on the surface of the ceiling facing the building interior or between the heated top living level and the unheated attic space;

Wall insulation, which is installed on the surface or in the cavity of an exterior wall;

Floor insulation, which is installed between the first level heated space of the dwelling and the unheated space beneath it, including a basement or crawl space;

Insulation for hot bare pipes, which is installed around the exterior of the pipes;

Roof insulation, which is placed on the surface of the roof facing the building interior or between a roof deck and its water repellent roof surface;

Exterior insulation for a hot water heater, which is placed around the exterior of the tank; and

Insulation for forced air ducts, which is wrapped around the exterior of the ducts.

Insulation does not include items that are primarily structural or decorative. For example, carpets, drapes, wood paneling, and exterior siding do not qualify although they may have been designed in part to have an insulating effect.

Storm or thermal windows include the following:
1. A window placed outside or inside an ordinary or prime window, creating an air space and providing greater resistance to heat flow and reduced air infiltration.

2. A window with enhanced resistance to heat flow through the glass area by multi-glazing, or with reduced air infiltration through weatherstripping. Multi-glazing is an arrangement in which two or more sheets of glazing material are fixed in a window frame to create one or more closed insulating spaces.

3. A window in which the glazed area consists of glass or other glazing materials with heat-absorbing or heat-reflecting properties that reduce the penetration of radiant heat through the window.

Storm or thermal doors include the following:
1. A second door, installed exterior to an existing outer

door, to provide greater resistance to heat flow and to reduce air infiltration.

2. A prime exterior door with enhanced resistance to heat flow through the door area because of reduced air infiltration through weather stripping and reduced heat flow through insulating wood or plastic frame material or metal material incorporating thermal breaks.

3. A glass door in which the glazed area is multi-glazed, or consists of glass or other materials with heat-absorbing or heat reflecting properties that reduce the penetration of radiant heat through the door.

Caulking consists of non-rigid materials placed in the joints of buildings to reduce the passage of air and moisture.

Weather stripping consists of narrow stips of flexible material placed over or in moveable joints of windows and doors to reduce the passage of air and moisture.

Automatic energy-saving setback thermostats are devices designed to reduce energy consumption by regulating the demand on the heating or cooling systems in which they are installed. These thermostats use a temperature control device for interior spaces incorporating more than one temperature control point, and a clock or other mechanism for switching from one control point to another.

Furnace replacement burners are devices for gas or oil-fired heating equipment that:

1. Mix the fuel with air and ignite the fuel-air mixture,

2. Are an integral part of a gas or oil-fired furnace or boiler, including the combustion chamber,

3. Replace an existing furnace burner, and

4. Are designed to achieve a reduction in the amount of fuel consumed as a result of increased combustion efficiency.

Flue opening modifications are dampers that, when installed in the pipe connecting the furnace to the chimney, conserve energy by substantially stopping the flow of air to the chimney when the furnace is not in operation.

Furnace ignition systems are electrical or mechanical devices that, when installed in a gas-fired heating system, ignite the fuel and replace a gas pilot light.

Items that do not qualify for the credit include heat pumps, fluorescent lights, wood- or peat-burning stoves, replacement boilers and furnaces, and hydrogen-fueled equipment.

Renewable Energy Source Expenditures

You may receive an additional energy credit for amounts you spend on solar, wind-powered, or geothermal property for your home. This credit is computed by taking 30% of the first $2,000 and 20% of the next $8,000 of these expenditures. You are eligible for the credit when the original installation of the property is completed. As in the case of energy conservation expenditures, the full $10,000 limit on renewable energy source expenditures may be spread over several tax years. A new $10,000 limit applies for each principal residence you occupy during the period of the credit.

Principal residence. Renewable energy source equipment, such as solar collectors, windmills, or geothermal wells, must be installed for use with your principal residence, which must be located in the United States. However, unlike the credit for energy conservation expenditures, the credit for renewable energy source equipment may be claimed for items installed for use with new, as well as existing, homes. For purposes of the credit it is immaterial when your home was constructed, as long as the renewable energy source property was installed after April 19, 1977.

However, if renewable energy source property is installed during construction or reconstruction of a residence, it is eligible for the credit when you first occupy the residence as your principal residence. Thus, if the property was installed before April 20, 1977, it may qualify for the credit if you first occupied the residence after April 19, 1977. But if you reoccupy a reconstructed dwelling that you had occupied as your principal residence before the reconstruction, the renewable energy source property is eligible for the credit when it is installed. "Reconstruction" is the replacement of most of a dwelling's major structures, such as floors, walls, and ceilings.

For purposes of the credit, a residence is considered your principal residence beginning 30 days before the date you cccupy it.

The renewable energy source property, to qualify, must be new, must be expected to last at least 5 years, and must meet certain performance and quality standards to be specified by the Secretary of the Treasury. As of the date of this publication, no performance and quality standards have been issued. However, the property does not need to meet the standards if it is purchased before the standards are published. The cost of renewable energy source equipment includes labor cost property allocable to the on-site preparation, assembly, or installation of the equipment.

Renewable energy source property includes the following:

Solar energy quipment for heating or cooling a dwelling or for providing hot water for use within the dwelling

Wind energy equipment for generating electricity or other forms of energy for personal residential purposes

Geothermal energy equipment

Additional devices, when specified by the Secretary of the Treasury, that rely on renewable energy sources for heating or cooling a dwelling or for providing hot water for use within the dwelling. As of the date of this publication, no additional devices have been so specified.

A renewable energy source expenditure does not include

any expenditure for a swimming pool used as an energy storage medium. Nor does it include an expenditure for an energy storage medium that has a primary function other than the function of energy storage. It also does not include an expenditure for a heating and cooling system to supplement renewable energy source equipment, if the supplementary system uses a form of energy other than solar, wind, or geothermal.

Solar energy property is equipment that uses solar energy to heat or cool a dwelling or to provide hot water for use within the dwelling. Generally, a solar energy system changes sunlight into heat or electricity through the use of equipment such as collectors (to absorb sunlight and create hot air), rockbeds (to store hot air), thermostats (to activate fans that circulate the hot air), and heat exchangers (to utilize hot air to heat water).

Solar energy property includes "passive" solar systems, "active" solar systems, and combinations of both types. An active solar system is based on the use of mechanically forced energy transfer, for example, using fans to circulate solar heat. A passive solar system is based on the use of conductive, convective, or radiant energy transfer, for example, using portions of the structure as solar furnaces to add heat to the structure. However, for purposes of the credit, materials and components that serve a significant structural function, or are structural components, such as extra-thick walls, windows, skylights, greenhouses, and roof overhangs, are not included as solar energy property.

Wind energy property is equipment that uses wind energy to produce energy in any form for personal residential purposes. Generally, wind energy equipment consists of a windmill that generates electricity and mechanical forms of energy. Equipment that uses wind energy for transportation does not qualify.

Geothermal energy property is equipment that uses geothermal energy to heat or cool a dwelling or to provide hot water for use within the dwelling. This is done by distributing or using geothermal deposits. A geothermal deposit is a geothermal reservoir containing natural heat stored in rocks, water, or vapor. For example, hot springs are a geothermal deposit.

How to claim the credit. You claim the credit on line 45 of your Form 1040 for 1978. The amount of credit is computed on Form 5695, *Energy Credits,* which you attach to your return. All eligible expenditures from April 20, 1977, through the end of 1978 are included in computing your credit for 1978. For this purpose, expenditures are generally treated as made when the original installation is completed. You may not amend a 1977 return to claim an energy credit for that year. You may not claim an energy credit on Form 1040A.

If your file your return on a fiscal year basis, you may not take the credit until your first tax year beginning after 1977.

Your residential energy credit must amount to at least $10 in any one year before it may be claimed. This $10 minimum applies *before* you consider the limitation to tax described in the following paragraph. The minimum applies to either joint or separate returns.

The credit may not exceed the amount of tax for which you are otherwise liable. Specifically, it is limited to the tax on line 37, Form 1040, minus the credits on lines 38 through 44, Form 1040.

Carryover of unused credit. You may carry over an unused credit (that you may not use because it exceeds your tax liability) to the next tax year. An unused credit may continue to be carried over to later tax years through 1987. For a fiscal year filer, an unused credit may be carried over through the fiscal year beginning in 1987.

Form **5695**

Department of the Treasury
Internal Revenue Service

Energy Credits

▶ Attach to Form 1040.

1978 ▶

Name(s) as shown on Form 1040.

Your social security number

Residential Energy Credit Computation

Energy Conservation Expenditures. For calendar year 1978 filers, energy conservation property must have been installed after April 19, 1977 and before January 1, 1979. For these expenditures to qualify for the credit, your principal residence must have been substantially completed before April 20, 1977.

(a) Description of Item (See Instruction B)	(b) Amount	
1 Insulation		
Storm (or thermal) windows or doors		
Caulking or weatherstripping		
Other (specify) ▶		
2 Total (add amounts on line 1) **2**		
3 Enter 15% of line 2 (but do not enter more than $300)		**3**

Renewable Energy Source Expenditures. For calendar year 1978 filers, renewable energy source property generally must have been installed after April 19, 1977 and before January 1, 1979.

(a) Description of Item (See Instruction C)	(b) Amount	
4 Solar		
Geothermal		
Wind		
5 Total (add amounts on line 4) **5**		
6 Enter 20% of line 5 (but do not enter more than $2,000) **6**		
7 Enter 10% of line 5 (but do not enter more than $200) **7**		
8 Total (add lines 3, 6, and 7—if less than $10, enter zero here and on line 10 below) . . **8**		
9 Limitation:		
a Enter tax from Form 1040, line 37 . . . . . . **9a**		
b Enter total of lines 38 through 44 from Form 1040 . . **9b**		
c Subtract line 9b from line 9a (if less than zero, enter zero) **9c**		
10 Residential energy credit. Enter the smaller of line 8 or line 9c here and on Form 1040, line 45 . . **10**		

ENERGY CREDITS

A. Who May Claim the Credit. Calendar year filers must file Form 5695 to claim a credit for energy saving property installed after April 19, 1977 and before January 1, 1979. Even if you installed an item in 1977 (after April 19), you must claim the credit on your 1978 return. Do not file an amended return for 1977.

Taxpayers with fiscal years beginning in 1977 and ending in 1978 may not claim the credit for the 1977-78 tax year. Fiscal year 1978-79 filers may claim the credit by taking into account the period beginning April 20, 1977 and ending on the last day of the tax year.

B. Energy Conservation Property. Items eligible for the credit are limited to the following:
  (1) insulation (fiberglass, cellulose, etc.) for: ceilings, walls, floors, roofs, water heaters, etc.;
  (2) exterior storm (or thermal) windows or doors;
  (3) caulking or weatherstripping for exterior windows or doors;
  (4) a furnace replacement burner which reduces the amount of fuel used;
  (5) a device to make flue openings (for a heating system) more efficient;
  (6) an electrical or mechanical furnace ignition system which replaces a gas pilot light;
  (7) an automatic energy-saving setback thermostat; and
  (8) a meter which displays the cost of energy usage.

These items must be installed in or on your principal residence (as defined in Instruction E) after April 19, 1977 and before January 1, 1979 and meet the following tests:
  (1) you must be the first person to use the item, and
  (2) the item can be expected to remain in use for at least 3 years.

C. Renewable Energy Source Property. Solar and geothermal energy property may be used to heat or cool your residence (or provide hot water). Solar energy property includes equipment (collectors, rockbeds, and heat exchangers) that transforms sunlight into heat or electricity.

Geothermal energy property includes equipment that distributes the natural heat in rocks or water. Wind energy property uses wind to produce energy in any form (generally electricity) for residential purposes.

Renewable energy source property must be installed in connection with your principal residence(as defined in Instruction E) and meet the following tests:
(1) you must be the first to use the item, and
(2) the item can be expected to remain in use for at least 5 years.

D. Items That Do Not Qualify for the Credit. Examples are:
(1) carpeting;
(2) drapes;
(3) wood paneling;
(4) exterior siding;
(5) heat pump;
(6) wood or peat fueled residential equipment;
(7) fluorescent replacement lighting system;
(8) hydrogen fueled residential equipment;
(9) equipment using wind energy for transportation;
(10) expenditures for a swimming pool used as an energy storage medium; and
(11) greenhouses.

E. Principal Residence Rules. The credit is available only for your principal residence (you may either own it or rent it from another person). It must be the main home occupied by you and your family. A summer or vacation home would not qualify. It must be located in the United States. To qualify for the energy conservation credit, the residence must have been substantially completed before April 20, 1977.

Note: Please get **Publication 903,** *Energy Credits for Individuals, for special rules about principal residences.*

F. Amount of Credit. The amount of the credit is based on the cost of the item. The cost of an energy conservation item includes its original installation. The cost of a renewable energy source item includes labor costs for its onsite preparation, assembly, or original installation.

The maximum credit for energy conservation items is $300 for each residence. The maximum credit for renewable energy source items is $2,200 for each residence.

Please get **Publication 903** for additional information if you occupied two or more principal residences and made expenditures for energy conservation property or renewable energy source property.

G. Unused Credit Carryover. If your energy credit for 1978 is more than your tax, you may carry over the unused amount to 1979.

Note: *For additional information, get* **Publication 903,** *Energy Credits for Individuals.*